Closing the Gap

CLOSING THE GAP

The Quest to Understand Prime Numbers

VICKY NEALE

OXFORD
UNIVERSITY PRESS

Great Clarendon Street, Oxford, OX2 6DP,
United Kingdom

Oxford University Press is a department of the University of Oxford.
It furthers the University's objective of excellence in research, scholarship,
and education by publishing worldwide. Oxford is a registered trade mark of
Oxford University Press in the UK and in certain other countries

First Edition published in 2017

Impression: 1

Published in the United States of America by Oxford University Press
198 Madison Avenue, New York, NY 10016, United States of America

British Library Cataloguing in Publication Data
Data available

Library of Congress Control Number: 2017948689

ISBN 978–0–19–878828–7

Printed and bound by
CPI Group (UK) Ltd, Croydon, CR0 4YY

CONTENTS

CHAPTER 1

Introduction

Until the nineteenth century, the Inaccessible Pinnacle was, well, inaccessible. A slender slice of basalt rising amongst the Cuillin mountains on the Isle of Skye, off the north-west coast of Scotland, it seemed unclimbable to the early pioneers of the nineteenth century who set out to explore the Cuillins. These days, it remains a challenging climb, but it is definitely not impossibly difficult: given fine weather, tour guides can take novices right to the top.

Mathematical exploration has much in common with this kind of adventuring. You stand looking at the sheer surface of your mathematical problem, searching for toeholds and crevices that might give a way up. After a long time looking, you start to make out an indistinct crack to the left, and a slight pattern in the rock up and to the right that reminds you of a climb you heard about once. Putting together all the features you've noticed, you can sketch out a possible route up the rock face, although it's not quite clear whether that small ledge will make a good toehold and there's a pretty ambitious reach near the top that might well be a stretch too far.

Still, now that you have a possible route in mind, you can start climbing, and hope that the details will become clearer along the way. Perhaps that reach will be too big, but when you get a bit closer maybe there'll be a crack in just the right place for your fingers.

Unfortunately, when you're three-quarters of the way up a sliver of rock breaks away, your toehold disappears from beneath your feet, and you drop back some way. Eventually, however, if you persevere you might reach the top.

Once someone has found their way to the top, suddenly the inaccessible becomes much more accessible. Once you know that *someone* has been up, you know that it can be done. If you have access to their notes, or know someone who heard them describe the route, then perhaps you can even follow in their footsteps. In some cases, what was at first a risky and demanding endeavour for pioneers becomes mainstream, and suitable for a weekend stroll even by those who have no climbing experience.

Now, there are many things about this that do not describe what it is like to do mathematics—and probably it's not such a great description of rock climbing either. But the analogy has its uses.

Rock climbing usually involves many people: often a challenge is taken on by a team rather than an individual, and also many teams will tackle the same rock face. In mathematics, there are romantic stories of individuals making heroic breakthroughs by themselves. Less well known are the collaborations, some of which in the twenty-first century involve very large numbers of people. This book has a romantic story of the best kind—an extraordinary breakthrough by an individual. It also has insights into these new large collaborations, and what they can reveal about what mathematicians do when they do mathematics. The Inaccessible Pinnacle for the mathematicians on this quest is the Twin Primes Conjecture, one of the most famous unsolved problems in the whole of mathematics. I'll tell you much more about it in the coming chapters.

I am no climber. If you are, then you will have detected this from the flaws in my description above! I do, however, love going on holiday to Skye, and when I am there I enjoy walking. I am inexperienced, and nervous about setting out on potentially hazardous trips by myself. However, I have had some wonderful days walking in the foothills of the Cuillins, seeking a personal challenge, enjoying the stunning scenery, catching glimpses of the peaks when they emerge from the clouds and marvelling at the skill, fitness and courage of those who reach the summits.

This book is for those who enjoy roaming the mathematical foothills. I hope to be the kind of guide that I would like for my trips to Skye: I want to show you the sights. In particular, I hope to give you glimpses of the summits, and to tell you tales of the people who climb them, while also leading you along my favourite routes in the foothills and pointing out some directions for adventures you might want to undertake in the future. Inevitably, this will present you with challenges. I have tried to select our routes carefully, and much of the journey is easy walking with spectacular views, but sometimes you will find that there is a tricky stretch, with some parts seeming out of reach. You have a big advantage over the hillwalker, though, because if a part is too challenging for you at the moment then you do not have to stop there: you can turn the page and skip over it in one bound! I'll try to highlight these parts, but I hope that you will skim through them to get some sense of the ideas involved, even if you do not want to go through them in detail on a first reading. This is, after all, how mathematicians usually read each other's papers and books.

With the scene set, it's time to set off. While you put on your mathematical walking boots, I'd like to thank some people.

I am very grateful to numerous friends and colleagues for their help with this book. Frances Kirwan encouraged me right from the start. Tim Gowers, Ben Green and James Maynard generously gave their time to tell me about work, both their own and that of others. Jennifer Balakrishnan, Rebecca Cotton-Barratt, Charlie Gilderdale, Lizzie Kimber, Ursula Martin, John Mason and Robin Wilson read drafts (and in some cases redrafts too), and gave me lots of constructive feedback and encouragement. Naturally the mistakes that remain are all my own. Keith Mansfield, then at Oxford University Press, supported me at the beginning of the project, and helped me to navigate the initial stages of the publishing process. Dan Taber, who took over the project from Keith, has been endlessly patient and kind, and came up with the title *Closing the Gap*.

In what follows, I mention many mathematicians of the past and present by name. Inevitably it has not been possible to name everyone whose work is relevant to this story, and I apologise in advance to those whose contributions I have not explicitly described. I would like to acknowledge their work and the progress that has been possible thanks to their involvement.

Although I had no idea of it at the time, my work on this project in a sense began when I was a Fellow at Murray Edwards College, Cambridge, when my colleagues invited me to give a Fellows' research talk and I had to find a way to describe ideas from additive number theory to a diverse and curious audience. I am grateful to my former colleagues at Murray Edwards for their support and friendship, and to my current colleagues at the Oxford Mathematical Institute and at Balliol College, Oxford for making me welcome in Oxford and for making it possible for me to write this book. The pencil that inspired Chapter 10 was kindly given to me by Paul Stephenson of The Magic Mathworks Travelling Circus, and I am very grateful for the idea. Some of the imagery and analogies in the chapters that follow arise from conversations with Charlie Gilderdale, and I am grateful to him for many discussions about mathematics and the teaching of mathematics.

My former and present undergraduate students, and the school students I have worked with over the years, are a constant source of inspiration.

Mathematical walking boots on? Let's set off on our quest to understand prime numbers!

CHAPTER 2

What is a prime?

I'd like to tell you some stories about prime numbers, so I should start by telling (or reminding) you what a prime number is. The prime numbers are the *indivisible* numbers. Let me draw you some pictures (Figure 2.1).

To draw the diagram for a particular number, we need to find a way to arrange the appropriate number of dots. We want to break down the number as much as possible. For example, 15 is 5 lots of 3 (that is, $15 = 5 \times 3$), and that's what we see in the picture. And 18 is 3 lots of 3 lots of 2 (that is, $18 = 3 \times 3 \times 2$), and that's what the diagram shows. There are other ways to break down 18, such as $18 = 3 \times 6$ and $18 = 2 \times 9$, but in each of those cases it's possible to subdivide further (since $6 = 2 \times 3$ and $9 = 3 \times 3$). But 2, 3 and 5 cannot be broken down further, they are indivisible, which is why we stop with those numbers.

The indivisible numbers can't be broken down at all, and we call these the *prime numbers*. They correspond to the rings of dots (except for 2, which is prime but doesn't quite have enough dots to form a ring). For example, there is no way that we can arrange 11 dots in groups of equal sizes. I've already mentioned that 2, 3 and 5 are indivisible, they're prime—that's why we had to stop with them when breaking up 15 and 18.

Every positive whole number bigger than 1 has at least two *factors*—that is, two numbers that exactly divide into it—namely 1 and the number itself. (Factors are also often called *divisors*.) Prime numbers are those numbers that have only these factors. For example, 13 is prime because its only factors are 1 and 13. By contrast, 12 is not prime, as we can see in Figure 2.1; 12 is divisible by 1 and 12, but has other factors too—for example, it's also divisible by 3 (and by 4) since $12 = 3 \times 4$.

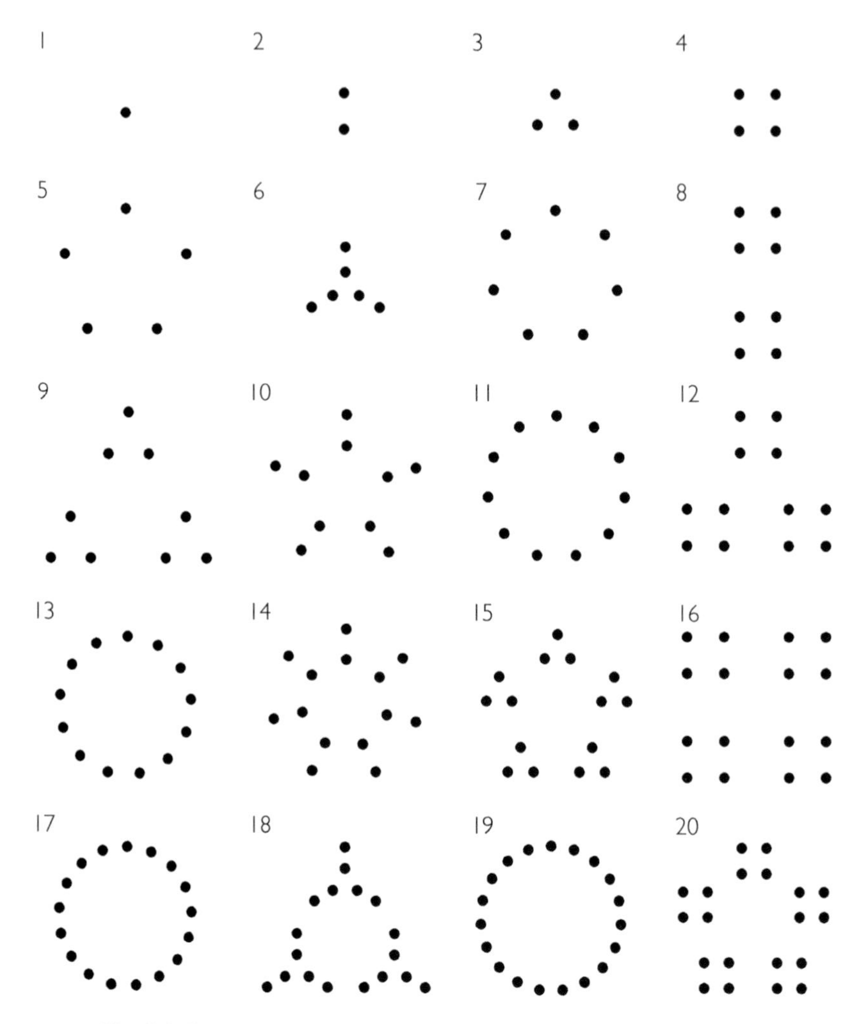

Fig. 2.1 Factorisations. Inspired by http://www.datapointed.net/visualizations/math/factorization/animated-diagrams/

In case you're wondering, I'll mention that 1 isn't a prime. That's not because of some interesting philosophical point, or because it's only got one number that divides into it rather than two, or anything like that. Definitions (such as that of a prime number) don't get handed to mathematicians on stone tablets. Rather, part of the job of mathematicians is to make good definitions, ones that lead to interesting mathematics, and it turns out to be better to define 1 not to be prime. I'll come back to that later.

Prime numbers are absolutely fundamental to mathematics, because they are the building blocks from which all whole numbers can be made. This is the idea captured by the diagrams we've just seen; I'll return to it in a bit more detail in Chapter 12. We'll see that to answer a question it is sometimes enough to answer it just for prime numbers, because we can then put that information together to answer it for all whole numbers.

Patterns in the primes

Let's have a good look at the primes (Figure 2.2).

1	2	3	4	5	6	7	8	9	10
11	12	13	14	15	16	17	18	19	20
21	22	23	24	25	26	27	28	29	30
31	32	33	34	35	36	37	38	39	40
41	42	43	44	45	46	47	48	49	50
51	52	53	54	55	56	57	58	59	60
61	62	63	64	65	66	67	68	69	70
71	72	73	74	75	76	77	78	79	80
81	82	83	84	85	86	87	88	89	90
91	92	93	94	95	96	97	98	99	100

Fig. 2.2 The primes up to 100.

I've displayed them in this grid because when I look at this arrangement I find myself noticing all sorts of interesting patterns. Crucially, the question for us as mathematicians is whether those patterns continue. If I'd included more rows, would we see the same patterns or were they just features of the first few rows?

For example, one thing I notice is that some of the columns seem to be empty. The columns with 4, 6, 8 and 10 at the top contain no primes, at least as far as we can see from this grid. Will that continue if we have more rows?

Well, if we think about it for a moment, then we realise that all the numbers in those columns are even, that is, that each of them is a multiple of 2. But an even number bigger than 2 can't possibly be a prime number: not only is it divisible by 1 and by itself, it's also divisible by 2. So we can be completely sure that those columns will never, ever contain any primes, no matter how many rows we include in the grid.

That's the power of mathematics. We can look at some data, notice an apparent pattern, make a conjecture (a conjecture is a statement of something that we think might be true), and then seek to prove it. And if we succeed in proving it, then we can be completely and utterly certain that it's true. I love that certainty.

We've just established that 2 is the only even prime number. In terms of the grid, this means that the only blue number in the 2, 4, 6, 8 and 10 columns is 2. We know that all the other entries in those columns (including the infinitely many numbers that we can't see because they are in rows not shown in the grid) are not blue (not prime).

I accept that's not the most exciting observation of all time. But it's a good example of a pattern that we might notice from the grid and then seek to explain.

In a similar spirit, we might notice that 5 seems to be the only blue number in its column. Does that pattern continue?

Well, each number in the 5 column is a multiple of 5, and apart from 5 every multiple of 5 is not prime (because it is divisible by itself, by 1 and also by 5). So indeed 5 is the only prime number in that column.

The other multiples of 5 all lie in the 10 column, so in particular are even. But this means that we've already established that they're not prime.

Can we use the same strategy for numbers other than 2 and 5? Where are the multiples of 3 in the grid, for example? This is shown in Figure 2.3.

1	2	3	4	5	6	7	8	9	10
11	12	13	14	15	16	17	18	19	20
21	22	23	24	25	26	27	28	29	30
31	32	33	34	35	36	37	38	39	40
41	42	43	44	45	46	47	48	49	50
51	52	53	54	55	56	57	58	59	60
61	62	63	64	65	66	67	68	69	70
71	72	73	74	75	76	77	78	79	80
81	82	83	84	85	86	87	88	89	90
91	92	93	94	95	96	97	98	99	100

Fig. 2.3 The multiples of 3 up to 100.

They form their own interesting patterns. When I look at the grid, I see diagonal lines from top right to bottom left. By the same sort of reasoning that we used for 2 and 5, we know that apart from 3 all these multiples of 3 cannot be prime (because they have 3 as a factor in addition to 1 and themselves). That immediately tells us something about the grid, but it's not quite as clear what it means as it was for multiples of 2 and 5. This relates to some ideas I'll explore further in Chapter 10.

What else might we ask ourselves?

One question that occurs to me is will we still keep getting more prime (blue) numbers if we continue the grid from Figure 2.2? It looks as though the primes are becoming more spread out towards the bottom of the grid—look at those big gaps from 73 to 79 to 83 to 89 to 97. A frog jumping from prime lily pad to prime lily pad is going to have to hop a long way to avoid the non-primes. Intuitively, that seems quite plausible—it's going to be hard for a really big number to be prime because there are so many smaller numbers that might divide into it.

So do we reach a biggest prime? There are two very different possibilities.

In one scenario, there's a biggest prime. After that prime, each number is divisible by some smaller number. Our grid stops having blue entries.

In the other scenario, there's no biggest prime. No matter how far down the grid we go, we keep finding blue entries (even if they're a bit thin on the ground). That is, there are infinitely many primes.

How do we decide? We could, with the help of a computer, draw up a much larger version of the grid, testing numbers up to a million, or a billion. What would that tell us? Absolutely nothing! Maybe we keep finding primes, maybe we don't. Either way, we still don't know what happens for all the numbers we *haven't* checked—and there are still infinitely many of them to go. The computer evidence might be useful for helping us to make a guess about whether or not there's a biggest prime, but it's never going to settle the matter one way or the other.

Fortunately, mathematicians have answered this question. The answer goes right back to Euclid, one of the most famous ancient Greek mathematicians. He was born around 325 BC and died around 265 BC, and in between he lived in Alexandria in Egypt and made massive contributions to mathematics. You might have come across Euclidean geometry, about which he wrote extensively in his series of books *The Elements*. This is the geometry we first meet at school—triangles, circles, Pythagoras's theorem, that sort of thing. But his books were not just about geometry, some of them were about number theory (that is, the study of properties of whole numbers). He included a proof that there are infinitely many primes, that is, that there is no biggest prime. So this is a *theorem*, something that mathematicians have proved to be true.

Theorem *There are infinitely many primes.*

Euclid's strategy was to use a technique called *proof by contradiction*. Here's an outline of his argument. We carry out a thought experiment. Secretly, we expect that there are infinitely many primes, but we imagine that we're in a universe in which there are only finitely many primes, and then see what that universe is like. It will turn out (in some way that will become clear later on) that a universe in which there are only finitely many primes would lead to something being both true and false at the same time. We call this a *contradiction*. Since this cannot happen (that is, we can't have a contradiction), that universe can't exist: there must be infinitely many primes. Now let's see the proof in a bit more detail.

Suppose that there are only finitely many primes. Then there is a largest prime, say p. We can write a list that contains all the primes in the world: 2, 3, 5, 7, ..., p. (If the largest prime in the world turned out to be 17, then the list would be 2, 3, 5, 7, 11, 13, 17.)

Here's the clever bit. We multiply all those primes together, and add 1. So we're considering the number $(2 \times 3 \times 5 \times 7 \times \cdots \times p) + 1$.

This is some very large number, we don't know exactly what it is because it involves the unknown largest prime p, but that's OK for our thought experiment.

One thing we do know about it is that it must have a prime factor, because every whole number greater than 1 has a prime factor. (Either it's prime itself, or it's divisible by a smaller prime.)

What can that prime factor be?

It can't be 2, because the number is $2 \times$ (something) + 1 and so leaves a remainder of 1 when we divide by 2.

And it can't be 3, because the number is $3 \times$ (something) + 1 and so leaves a remainder of 1 when we divide by 3.

And it can't be 5, because . . . well, you get the idea. All the way up to p.

We've built a number $(2 \times 3 \times 5 \times 7 \times \cdots \times p) + 1$ that has a prime factor, but that isn't divisible by any of the primes on our list.

But that list was supposed to contain all the primes in the world!

We've reached a *contradiction*: we have found a number that is both divisible by a prime and yet not divisible by any of the primes in the world.

This tells us that our initial supposition, that there are only finitely many primes, cannot be correct. So there must be infinitely many primes. $\square$

(The little box is the symbol that mathematicians use to indicate the end of a proof. In the past, people sometimes put QED at the end of a proof. It's an abbreviation for *quod erat demonstrandum*, which translates from Latin as something like *that which was to be proved*. I like to think of the little box as a less pompous equivalent!)

A small note of caution: we've shown that the number $(2 \times 3 \times 5 \times 7 \times \cdots \times p) + 1$ isn't divisible by $2, 3, 5, 7, \ldots, p$, but that doesn't mean that it's necessarily prime. It might be prime—for example, $(2 \times 3 \times 5 \times 7 \times 11) + 1 = 2311$ is prime. But $(2 \times 3 \times 5 \times 7 \times 11 \times 13) + 1 = 30031 = 59 \times 509$ is not prime. It can be tempting to believe that the number we construct in the proof is always prime, but that really isn't the case. It doesn't matter for Euclid's argument, all that matters there is that this number isn't divisible by any of the primes we started with.

I love Euclid's proof. I love the fact that we have something so difficult to prove (that there are infinitely many primes), and it seems like we're never going to be able to prove it (because just finding more and more primes will never be enough), and then Euclid finds a way to prove it with two very clever and elegant ideas—proof by contradiction, and using a finite list of primes to build a new number not divisible by any of them. It's the sort of proof that would be hard to think of, but when someone tells you about it, and you have time to think about it, starts to feel natural. It also gives us a strategy for proving other results with a similar flavour (that there are infinitely many primes with some particular property, for example). Mathematicians do this all the time—having had a good idea, we try to exploit it as much as possible. More on that later ...

Gaps between primes

We now know that there is no largest prime: no matter how far we extend the grid, we'll keep finding new prime numbers. However, it might still be the case that they become rather spread out as we go further down the grid. Our intuition earlier was that it's hard for a large number to be prime, because there are many smaller numbers that might divide it.

But it turns out that it's a bit more subtle than that. This is a recurring feature of the prime numbers: every time we think we understand them, we find there's more to them than we realised! I find this enormously intriguing, it's one of the things that makes the study of prime numbers so compelling.

In this case, by including only the numbers up to 100 I'm misleading you somewhat. If I include an extra row, the situation looks a bit different (Figure 2.4).

1	2	3	4	5	6	7	8	9	10
11	12	13	14	15	16	17	18	19	20
21	22	23	24	25	26	27	28	29	30
31	32	33	34	35	36	37	38	39	40
41	42	43	44	45	46	47	48	49	50
51	52	53	54	55	56	57	58	59	60
61	62	63	64	65	66	67	68	69	70
71	72	73	74	75	76	77	78	79	80
81	82	83	84	85	86	87	88	89	90
91	92	93	94	95	96	97	98	99	100
101	102	103	104	105	106	107	108	109	110

Fig. 2.4 The primes up to 110.

Ah. The numbers 101, 103, 107 and 109 are all prime, and they're pretty bunched up.

What happens if I include lots more rows? If we zoom out of the grid, so that I can fit lots more numbers on the page, we can get a sense of that (Figure 2.5).

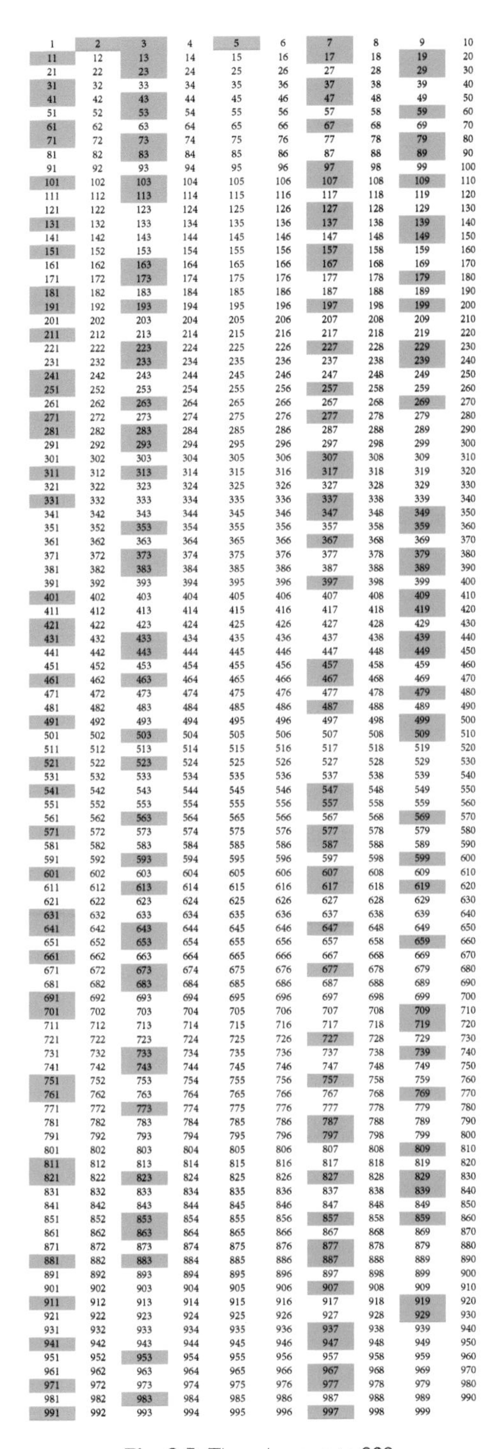

1	**2**	**3**	4	**5**	6	**7**	8	9	10
11	12	**13**	14	15	16	**17**	18	**19**	20
21	22	**23**	24	25	26	27	28	**29**	30
31	32	33	34	35	36	**37**	38	39	40
41	42	**43**	44	45	46	**47**	48	49	50
51	52	**53**	54	55	56	57	58	**59**	60
61	62	63	64	65	66	**67**	68	69	70
71	72	**73**	74	75	76	77	78	**79**	80
81	82	**83**	84	85	86	87	88	**89**	90
91	92	93	94	95	96	**97**	98	99	100
101	102	**103**	104	105	106	**107**	108	**109**	110
111	112	**113**	114	115	116	117	118	119	120
121	122	123	124	125	126	**127**	128	129	130
131	132	133	134	135	136	**137**	138	**139**	140
141	142	143	144	145	146	147	148	**149**	150
151	152	153	154	155	156	**157**	158	159	160
161	162	**163**	164	165	166	**167**	168	169	170
171	172	**173**	174	175	176	177	178	**179**	180
181	182	183	184	185	186	187	188	189	190
191	192	**193**	194	195	196	**197**	198	**199**	200
201	202	203	204	205	206	207	208	209	210
211	212	213	214	215	216	217	218	219	220
221	222	**223**	224	225	226	**227**	228	**229**	230
231	232	**233**	234	235	236	237	238	**239**	240
241	242	243	244	245	246	247	248	249	250
251	252	253	254	255	256	**257**	258	259	260
261	262	**263**	264	265	266	267	268	**269**	270
271	272	273	274	275	276	**277**	278	279	280
281	282	**283**	284	285	286	287	288	289	290
291	292	**293**	294	295	296	297	298	299	300
301	302	303	304	305	306	**307**	308	309	310
311	312	**313**	314	315	316	**317**	318	319	320
321	322	323	324	325	326	327	328	329	330
331	332	333	334	335	336	**337**	338	339	340
341	342	343	344	345	346	**347**	348	**349**	350
351	352	**353**	354	355	356	357	358	**359**	360
361	362	363	364	365	366	**367**	368	369	370
371	372	**373**	374	375	376	377	378	**379**	380
381	382	**383**	384	385	386	387	388	**389**	390
391	392	393	394	395	396	**397**	398	399	400
401	402	403	404	405	406	407	408	**409**	410
411	412	413	414	415	416	417	418	**419**	420
421	422	423	424	425	426	427	428	429	430
431	432	**433**	434	435	436	437	438	**439**	440
441	442	**443**	444	445	446	447	448	**449**	450
451	452	453	454	455	456	**457**	458	459	460
461	462	**463**	464	465	466	**467**	468	469	470
471	472	473	474	475	476	477	478	**479**	480
481	482	483	484	485	486	**487**	488	489	490
491	492	493	494	495	496	497	498	**499**	500
501	502	**503**	504	505	506	507	508	**509**	510
511	512	513	514	515	516	517	518	519	520
521	522	**523**	524	525	526	527	528	529	530
531	532	533	534	535	536	537	538	539	540
541	542	543	544	545	546	**547**	548	549	550
551	552	553	554	555	556	**557**	558	559	560
561	562	**563**	564	565	566	567	568	**569**	570
571	572	573	574	575	576	**577**	578	579	580
581	582	583	584	585	586	**587**	588	589	590
591	592	**593**	594	595	596	597	598	**599**	600
601	602	603	604	605	606	**607**	608	609	610
611	612	**613**	614	615	616	**617**	618	**619**	620
621	622	623	624	625	626	627	628	629	630
631	632	633	634	635	636	637	638	639	640
641	642	**643**	644	645	646	**647**	648	649	650
651	652	**653**	654	655	656	657	658	**659**	660
661	662	663	664	665	666	667	668	669	670
671	672	**673**	674	675	676	**677**	678	679	680
681	682	**683**	684	685	686	687	688	689	690
691	692	693	694	695	696	697	698	699	700
701	702	703	704	705	706	707	708	**709**	710
711	712	713	714	715	716	717	718	**719**	720
721	722	723	724	725	726	**727**	728	729	730
731	732	**733**	734	735	736	737	738	**739**	740
741	742	**743**	744	745	746	747	748	749	750
751	752	753	754	755	756	**757**	758	759	760
761	762	763	764	765	766	767	768	**769**	770
771	772	**773**	774	775	776	777	778	779	780
781	782	783	784	785	786	**787**	788	789	790
791	792	793	794	795	796	**797**	798	799	800
801	802	803	804	805	806	807	808	**809**	810
811	812	813	814	815	816	817	818	819	820
821	822	**823**	824	825	826	**827**	828	**829**	830
831	832	833	834	835	836	837	838	**839**	840
841	842	843	844	845	846	847	848	849	850
851	852	**853**	854	855	856	**857**	858	**859**	860
861	862	**863**	864	865	866	867	868	869	870
871	872	873	874	875	876	**877**	878	879	880
881	882	**883**	884	885	886	**887**	888	889	890
891	892	893	894	895	896	897	898	899	900
901	902	903	904	905	906	**907**	908	909	910
911	912	913	914	915	916	917	918	**919**	920
921	922	923	924	925	926	927	928	**929**	930
931	932	933	934	935	936	**937**	938	939	940
941	942	943	944	945	946	**947**	948	949	950
951	952	**953**	954	955	956	957	958	959	960
961	962	963	964	965	966	**967**	968	969	970
971	972	973	974	975	976	**977**	978	979	980
981	982	**983**	984	985	986	987	988	989	990
991	992	993	994	995	996	**997**	998	999	

Fig. 2.5 The primes up to 999.

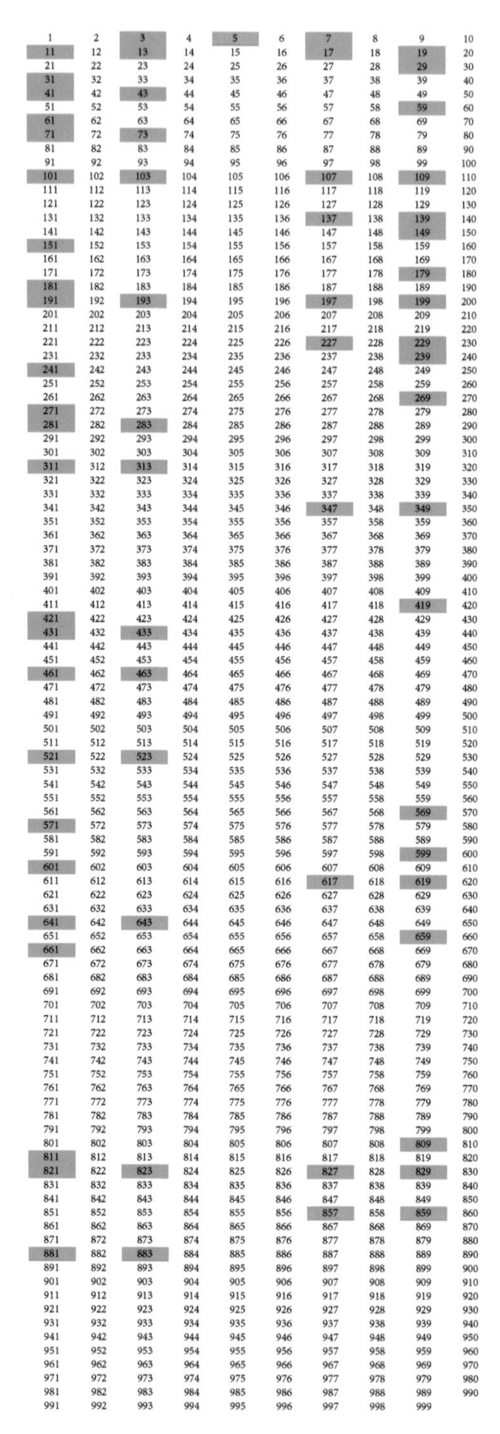

1	2	**3**	4	**5**	6	**7**	8	9	10
11	12	**13**	14	15	16	**17**	18	**19**	20
21	22	23	24	25	26	27	28	**29**	30
31	32	33	34	35	36	37	38	39	40
41	42	**43**	44	45	46	47	48	49	50
51	52	53	54	55	56	57	58	**59**	60
61	62	63	64	65	66	67	68	69	70
71	72	**73**	74	75	76	77	78	79	80
81	82	83	84	85	86	87	88	89	90
91	92	93	94	95	96	97	98	99	100
101	102	**103**	104	105	106	**107**	108	**109**	110
111	112	113	114	115	116	117	118	119	120
121	122	123	124	125	126	127	128	129	130
131	132	133	134	135	136	**137**	138	**139**	140
141	142	143	144	145	146	147	148	**149**	150
151	152	153	154	155	156	157	158	159	160
161	162	163	164	165	166	167	168	169	170
171	172	173	174	175	176	177	178	**179**	180
181	182	183	184	185	186	187	188	189	190
191	192	**193**	194	195	196	**197**	198	**199**	200
201	202	203	204	205	206	207	208	209	210
211	212	213	214	215	216	217	218	219	220
221	222	223	224	225	226	**227**	228	**229**	230
231	232	233	234	235	236	237	238	**239**	240
241	242	243	244	245	246	247	248	249	250
251	252	253	254	255	256	257	258	259	260
261	262	263	264	265	266	267	268	**269**	270
271	272	273	274	275	276	277	278	279	280
281	282	**283**	284	285	286	287	288	289	290
291	292	293	294	295	296	297	298	299	300
301	302	303	304	305	306	307	308	309	310
311	312	**313**	314	315	316	317	318	319	320
321	322	323	324	325	326	327	328	329	330
331	332	333	334	335	336	337	338	339	340
341	342	343	344	345	346	**347**	348	**349**	350
351	352	353	354	355	356	357	358	359	360
361	362	363	364	365	366	367	368	369	370
371	372	373	374	375	376	377	378	379	380
381	382	383	384	385	386	387	388	389	390
391	392	393	394	395	396	397	398	399	400
401	402	403	404	405	406	407	408	409	410
411	412	413	414	415	416	417	418	**419**	420
421	422	423	424	425	426	427	428	429	430
431	432	**433**	434	435	436	437	438	439	440
441	442	443	444	445	446	447	448	449	450
451	452	453	454	455	456	457	458	459	460
461	462	**463**	464	465	466	467	468	469	470
471	472	473	474	475	476	477	478	479	480
481	482	483	484	485	486	487	488	489	490
491	492	493	494	495	496	497	498	499	500
501	502	503	504	505	506	507	508	509	510
511	512	513	514	515	516	517	518	519	520
521	522	**523**	524	525	526	527	528	529	530
531	532	533	534	535	536	537	538	539	540
541	542	543	544	545	546	547	548	549	550
551	552	553	554	555	556	557	558	559	560
561	562	563	564	565	566	567	568	**569**	570
571	572	573	574	575	576	577	578	579	580
581	582	583	584	585	586	587	588	589	590
591	592	593	594	595	596	597	598	**599**	600
601	602	603	604	605	606	607	608	609	610
611	612	613	614	615	616	**617**	618	**619**	620
621	622	623	624	625	626	627	628	629	630
631	632	633	634	635	636	637	638	639	640
641	642	**643**	644	645	646	647	648	649	650
651	652	653	654	655	656	657	658	**659**	660
661	662	663	664	665	666	667	668	669	670
671	672	673	674	675	676	677	678	679	680
681	682	683	684	685	686	687	688	689	690
691	692	693	694	695	696	697	698	699	700
701	702	703	704	705	706	707	708	709	710
711	712	713	714	715	716	717	718	719	720
721	722	723	724	725	726	727	728	729	730
731	732	733	734	735	736	737	738	739	740
741	742	743	744	745	746	747	748	749	750
751	752	753	754	755	756	757	758	759	760
761	762	763	764	765	766	767	768	769	770
771	772	773	774	775	776	777	778	779	780
781	782	783	784	785	786	787	788	789	790
791	792	793	794	795	796	797	798	799	800
801	802	803	804	805	806	807	808	**809**	810
811	812	813	814	815	816	817	818	819	820
821	822	**823**	824	825	826	**827**	828	**829**	830
831	832	833	834	835	836	837	838	839	840
841	842	843	844	845	846	847	848	849	850
851	852	853	854	855	856	**857**	858	**859**	860
861	862	863	864	865	866	867	868	869	870
871	872	873	874	875	876	877	878	879	880
881	882	**883**	884	885	886	887	888	889	890
891	892	893	894	895	896	897	898	899	900
901	902	903	904	905	906	907	908	909	910
911	912	913	914	915	916	917	918	919	920
921	922	923	924	925	926	927	928	929	930
931	932	933	934	935	936	937	938	939	940
941	942	943	944	945	946	947	948	949	950
951	952	953	954	955	956	957	958	959	960
961	962	963	964	965	966	967	968	969	970
971	972	973	974	975	976	977	978	979	980
981	982	983	984	985	986	987	988	989	990
991	992	993	994	995	996	997	998	999	

Fig. 2.6 The twin primes up to 999.

On this zoomed-out grid, it still looks as though on the whole the primes are becoming more spread out, but there continue to be some primes that are very close together, such as 347 and 349, and 659 and 661, and even the little cluster 821, 823, 827 and 829.

It seems reasonable to wonder whether we keep getting occasional instances of primes that are bunched up, even if on average they're becoming more sparse. For example, are there infinitely many pairs of primes that differ by just 2? Pairs like 3 and 5, or 17 and 19, or 101 and 103, highlighted in Figure 2.6. Such pairs of primes that differ by 2 are called *twin primes*.

I like to imagine that Euclid asked himself whether there are infinitely many pairs of twin primes, because it seems such a natural thing to wonder, especially having just proved that there are infinitely many primes. Unfortunately, I don't know the answer to this question …

CHAPTER 3

May 2013

Are there infinitely many pairs of primes that differ by just 2? I don't know—but neither does anyone else. In fact, this is an unsolved problem. As I wrote that sentence, I was acutely aware that it could become untrue at any moment, if someone solves the problem. That's what it's like at the cutting edge of mathematical research.

But in May 2013, common rooms in mathematics departments across the world were buzzing with news. A little-known mathematician had made huge progress on this problem. He spoke about his work at a Harvard seminar on 13th May 2013, and mathematicians immediately started spreading the news via excited emails and blog posts. Yitang Zhang, a lecturer at the University of New Hampshire, had proved that there are infinitely many pairs of primes that differ by at most 70 000 000. To put that another way, there are infinitely many pairs of prime numbers with a gap that is less than or equal to 70 million.

Zhang's work was a big leap towards the famous *Twin Primes Conjecture*, which asserts that there are infinitely many pairs of primes that differ by 2, pairs like 3 and 5, or 17 and 19, or 101 and 103, those pairs we were thinking about at the end of the last chapter. I'll say more about the Twin Primes Conjecture in the next chapter.

If mathematicians are aiming to show that there are infinitely many pairs of primes that differ by 2, why the excitement about Zhang's result? After all, 70 000 000 is an extremely large number, especially when we're aiming for 2.

Well, this was the first time anyone had managed to prove *any* result of this form. And while 70 million is an enormous number, it's still finite, and that's a huge step forwards towards a solution to the problem.

In April 2013, Zhang had submitted his paper 'Bounded gaps between primes' to the *Annals of Mathematics*, one of the most prestigious journals in the subject. The editors of the *Annals* are used to receiving papers containing ambitious claims from unknown mathematicians. Mathematics journals and departments, and individual mathematicians, frequently receive messages from people claiming to have solved famous problems. Unfortunately, the arguments are very often flawed. It's a bad sign when someone claims to have done something that mathematicians already know is impossible, for example, but it happens all the time. This paper was different. It was written very clearly, and showed a deep understanding of the work of the last couple of decades by many of the experts in the area—indeed, it built on their results. If Zhang's argument was correct, then the work would be ground-breaking. The editors accelerated the usual peer-review process, which often takes months, and on 21st May 2013, following enthusiastic reports from referees (experts in the area who had checked the paper carefully) and small modifications from Zhang, the paper was accepted for publication by the *Annals* (and published online on 1st May 2014).

Part of the excitement in the mathematics common rooms was because Zhang was not renowned for his work in the subject. In fact, he had struggled to find an academic job following his PhD, which he received from Purdue University in 1991. His doctorate was in pure mathematics, but not on gaps between prime numbers. He spent several years taking jobs outside academia, including a spell working for the Subway restaurant chain, before becoming a lecturer at the University of New Hampshire, a role with a large teaching load. Throughout this time, he kept up with developments in the mathematical community, working very much by himself, and so read about the work that would turn out to form the foundations for his own breakthrough. By 2013, Zhang was in his late 50s, so he confounds the stereotype that brilliant theorems are proved by young people at the starts of their careers.

I'm being very careful to say that Zhang was claiming that there are infinitely many pairs of primes that differ by *at most* 70 000 000. We can deduce from this that there's some number, say k, with $k \leqslant 70\,000\,000$, that has the property that there are infinitely many primes p so that $p + k$ is also prime. We hope that 2 has this property, we expect that in fact many numbers less than 70 million have this property. What Zhang's work tells us is that there's at least one such number less than or equal to 70 million.

All this has suddenly transformed Zhang from a relatively unknown lecturer to a mathematical celebrity. He was promoted to Professor at the University of

New Hampshire, and has since moved to become Professor at the University of California, Santa Barbara. In addition, he has received a shower of awards, for example the 2013 Ostrowski Prize, which is 'awarded every other year to a mathematician or to a group of scientists who have produced the best result in the field of pure mathematics or in the foundations of numerical mathematics'. In 2014 he received a MacArthur Fellowship—these five-year unrestricted grants are given 'to talented individuals who have shown extraordinary originality and dedication in their creative pursuits and a marked capacity for self-direction'. Every four years, the mathematical community gathers at the International Congress of Mathematicians (ICM), held in a different location round the world each time. In 2014, the ICM was in Seoul, and Zhang was one of the invited speakers. While this may not sound much, it is regarded as a great honour for a mathematician to be invited to speak at an ICM. It is not clear that Zhang welcomes the media attention arising from his spectacular work, and certainly, after so many years of working in relative obscurity, it is a big change for him in his late 50s.

Building on the work of numerous other mathematicians, but also adding ideas of his own and displaying great technical perseverance to make it all work, Zhang had shown that there are infinitely many pairs of primes that differ by at most 70 000 000. And then the game was on: could mathematicians close the gap, by making that number smaller?

CHAPTER 4

It's easy to ask hard questions

The Twin Primes Conjecture is one of the most famous unsolved problems in mathematics. The conjecture predicts that there are infinitely many pairs of twin primes, pairs of prime numbers that differ by 2, like 3 and 5, or 17 and 19, or 101 and 103. To say that in another way, the Twin Primes Conjecture asserts that there are infinitely many primes p such that $p+2$ is also prime. The challenge for mathematicians is to prove the conjecture—or to disprove it. It's not clear who first made the conjecture: it could go right back to Euclid and the ancient Greeks, or it might be more recent, I just don't know. In 1849, Alphonse de Polignac (1826–1863) made a more general conjecture of which the Twin Primes Conjecture is a special case, so it certainly goes back that far, but it seems likely that it is much older.

There are some pretty good reasons to think that the Twin Primes Conjecture might be true. One of them is that, with the help of computers, people have found many pairs of twin primes, including some very large numbers indeed—there is a known pair of twin primes each with more than 200 000 digits. But, just as with primes themselves, there is no prospect of *proving* that there are infinitely many pairs of twin primes by finding bigger and bigger examples, so the computations to find these numbers are a bit of a sideline rather than a major contribution to the mathematical quest.

The Twin Primes Conjecture has the same sort of flavour as a theorem that we saw in Chapter 2, the one that says that there are infinitely many primes. The difference is that we already know how to prove that there are infinitely many primes, thanks to Euclid. Maybe we can use some of the ideas from his argument to help us.

Let's play around with an example.

If we start with the primes 2, 3 and 5, we can use the strategy from Euclid's proof to create a number that is not divisible by 2, 3 or 5. We multiply them all together and add 1, to get $(2 \times 3 \times 5) + 1 = 31$. And a quick check (or flick back to our earlier grid of primes) shows that 31 is prime.

If instead of adding 1 we subtract 1, then we get $(2 \times 3 \times 5) - 1 = 29$, and again this is prime. (Euclid could just as well have subtracted 1, rather than adding it, in his proof that there are infinitely many primes, that would would have worked too.) And now we have a pair of twin primes: 29 and 31, which differ by 2.

What happens if we try the same thing more generally?

Let's start with the primes 2, 3, 5, ..., p, and multiply them all together and add 1, which gives the number $(2 \times 3 \times 5 \times \cdots \times p) + 1$. The point about this new number is that it's not divisible by any of 2, 3, 5, ..., p. And we can also multiply all these primes and *subtract* 1, which gives the number $(2 \times 3 \times 5 \times \cdots \times p) - 1$, which is also not divisible by any of 2, 3, 5, ..., p, for just the same reason. So now we have two numbers that differ by 2 and that aren't divisible by any of the primes up to p.

Why does this not immediately give us infinitely many pairs of twin primes?

The problem is that there's no guarantee that these new numbers we build will be prime, all we know is that they're not divisible by any of 2, 3, 5, ..., p. (I mentioned this at the end of Euclid's proof that there are infinitely many primes.) Let me show you what I mean with an example.

If we start with the primes 2, 3, 5 and 7, then we want to consider the two numbers $(2 \times 3 \times 5 \times 7) + 1 = 211$ and $(2 \times 3 \times 5 \times 7) - 1 = 209$. Now 211 is prime, but 209 isn't: it factorises as $209 = 11 \times 19$.

The problem is that the number we construct from the primes 2, 3, 5, ..., p doesn't have to be prime—it can't be divisible by any of 2, 3, 5, ..., p, but it absolutely *can* be divisible by a prime larger than p. This is what happened in the example in the last paragraph: we used the primes 2, 3, 5 and 7 to construct the number 209, which is divisible by the primes 11 and 19.

It seemed a moment ago that we might have a way to construct infinitely many pairs of primes that differ by 2, but unfortunately our idea turned out not to go anywhere. This happens a lot in mathematics. The American mathematician Julia Robinson (1919–1985), one of my mathematical heroes, was once asked to describe what she did each day. Her description was

> 'Monday—tried to prove theorem, Tuesday—tried to prove theorem, Wednesday—tried to prove theorem, Thursday—tried to prove theorem, Friday—theorem false'.

Even when the theorem is true, mathematicians end up trying many strategies for proving it that turn out not to work. Sometimes one can learn a lot from unpicking why an approach is unsuccessful, sometimes one just has to accept it and try something else instead. Mathematicians have to be resilient: the normal state for a mathematician is being stuck on a problem (or on several problems). After all, if you solve a problem then you move on to the next one and get stuck on that!

Goldbach's Conjecture

The Twin Primes Conjecture isn't the only example of a question about prime numbers that's easy to state but that turns out to be extremely hard to answer.

Instead of looking at gaps between primes, let's do something slightly different. Let's try adding primes. Which numbers can we write as the sum of two primes? For example, $10 = 3 + 7$, and 3 and 7 are both prime, so 10 is a sum of two primes. Similarly, $8 = 3 + 5$ and $14 = 3 + 11$, and so on.

I thought I'd draw up another grid, showing the numbers that are sums of two primes. Actually, I'm just going to focus on odd primes here: I'm going to ignore 2, which is a special case because it's the only even prime. It turns out to be more interesting just with odd primes. So here's the first step in my working: I wrote out a list of primes, and added 3 (the first odd prime) to each of them, and I shaded the results on my grid (Figure 4.1).

1	2	3	4	5	6	7	8	9	10
11	12	13	14	15	16	17	18	19	20
21	22	23	24	25	26	27	28	29	30
31	32	33	34	35	36	37	38	39	40
41	42	43	44	45	46	47	48	49	50
51	52	53	54	55	56	57	58	59	60
61	62	63	64	65	66	67	68	69	70
71	72	73	74	75	76	77	78	79	80
81	82	83	84	85	86	87	88	89	90
91	92	93	94	95	96	97	98	99	100

Fig. 4.1 Odd primes plus 3.

Then I added 5 (the next odd prime) to each prime, and coloured in the results. Some of them were already shaded, but that doesn't matter, I didn't try to shade them again. For example, 16 was already highlighted because 16 = 3 + 13, so I didn't colour it again, even though 16 = 5 + 11. Then I added 7, and 11, and all the odd primes until the answers were too big for my grid. You might like to pause to do this for yourself, to see what you get. Figure 4.2 shows what I came up with.

1	2	3	4	5	6	7	8	9	10
11	12	13	14	15	16	17	18	19	20
21	22	23	24	25	26	27	28	29	30
31	32	33	34	35	36	37	38	39	40
41	42	43	44	45	46	47	48	49	50
51	52	53	54	55	56	57	58	59	60
61	62	63	64	65	66	67	68	69	70
71	72	73	74	75	76	77	78	79	80
81	82	83	84	85	86	87	88	89	90
91	92	93	94	95	96	97	98	99	100

Fig. 4.2 Sums of two odd primes.

That's a pretty striking pattern, isn't it?

My eye is drawn to two things. The first is that the 1, 3, 5, 7 and 9 columns seem to be empty: they appear to contain no sums of two odd primes. The second is that apart from 2 and 4 themselves, the 2, 4, 6, 8 and 10 columns seem to be full—it looks as though every entry is a sum of two odd primes.

Let's think about that a bit. What happens if we add two odd primes? Actually, what happens if we add any two odd numbers (regardless of whether they're prime)? If we try some examples and then think about it a bit, we realise that we must get an even number. If you haven't thought about that before, then maybe you want to take a moment to convince yourself that it really is true. You could use algebra, or you could draw pictures of bars of chocolate (Figure 4.3).

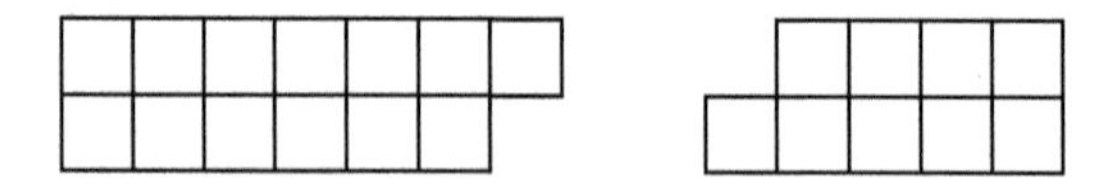

Fig. 4.3 The sum of two odd numbers is an even number.

In particular, no odd number is ever a sum of two odd primes. So, with hindsight, I'm not at all surprised that the 1, 3, 5, 7 and 9 columns are empty, and moreover I'm sure that they are completely empty, whereas before I only knew that the entries in the first few rows were empty.

What about the even columns? It's certainly possible for an even number to be a sum of two odd primes, but that doesn't mean that they all are. The grid records my calculations which checked that every even number from 6 to 100 inclusive is a sum of two odd primes. Does that continue if I add more rows to the grid?

You may by now have guessed how this works. I present to you some intriguingly simple question, and then say that not only do I not know the answer, but neither does anyone else. This is indeed such a problem. This one is called *Goldbach's Conjecture*. Christian Goldbach (1690–1764) was a mathematician now known for one thing, this conjecture, which he made in a letter to the signficantly more famous mathematician Leonhard Euler, in 1742. Euler (1703–1783) and Goldbach both worked as mathematicians in Russia, variously in St Petersburg and Moscow, and in 1729 they began a correspondence that lasted for many years, much of it about number theory. I am intrigued to find that their letters were written in a mix of German and Latin, and the way in which they talked about mathematics is slightly different from the way in which we phrase things now (they used different terminology), so I am not going to quote Goldbach directly. When we talk about 'Goldbach's Conjecture' today, we mean the assertion that every even number greater than 4 is a sum of two odd primes. The fact that this problem is known as Goldbach's Conjecture doesn't necessarily mean that he was the first to state the conjecture—the history of mathematics is littered with results and conjectures named after the wrong people. At least in this case there is evidence that Goldbach did actually make the conjecture!

As with the Twin Primes Conjecture, computers can check many cases of Goldbach's Conjecture, and so far the evidence is good: we haven't found a counterexample (an even number that cannot be written as a sum of two primes). But, again as with the Twin Primes Conjecture, that doesn't resolve the problem, because a computer can only ever check finitely many cases, leaving infinitely many unchecked cases. Until mathematicians find a proof, the conjecture remains open.

Germain primes

Let me tell you about one more irresistible unsolved problem before I move on.

Another of my mathematical heroes is Monsieur LeBlanc, a mathematician who worked in France at the end of the eighteenth and start of the nineteenth centuries. LeBlanc contributed to various fields, including trying to formulate a mathematical theory of elasticity, but for me the most significant contribution was to number theory. LeBlanc was a pseudonym, adopted by the French mathematician Sophie Germain (1776–1831) in an era when it was not socially acceptable for women to study mathematics.

When she was a teenager, her parents were not keen for her to study, but she was resolved to become a mathematician and so studied under the blankets at night. The École Polytechnique was founded in Paris just at the right time for Germain—but she was not allowed to study there because of her gender. Nonetheless, she managed to obtain many of the lecture notes, and first adopted the pseudonym M. LeBlanc in order to conceal her true identity from the great mathematician Joseph-Louis Lagrange when sending him some work having completed his university course.

She went on to correspond with other noted figures, of whom perhaps the most significant was Carl Friedrich Gauss. Again, she took the pseudonym Monsieur LeBlanc out of anxiety that he would not take her seriously if he knew her true identity. Gauss did eventually discover that M. LeBlanc was in fact Sophie Germain, following a somewhat surprising sequence of events. The young Sophie had been motivated to become a mathematician in part because she had read about the death of Archimedes, who was killed by a Roman soldier, apparently while he was working on a piece of mathematics. When, in 1806, the French occupied Gauss's hometown, Germain remembered the story of Archimedes and became concerned that Gauss might be in danger, so wrote to a family friend who was a senior figure in the French military, asking him to intervene. This led to Gauss learning that M. LeBlanc and Sophie Germain were one and the same, and to his credit, his response was to offer her further praise for her work rather than refusing to discuss mathematics with a woman.

Germain is remembered today chiefly for her work on Fermat's Last Theorem. That's a story for another time, and I'll say more about it in Chapter 12, but it is relevant here because in her work on the Fermat problem Germain found herself studying primes p with the property that $2p + 1$ is also prime. Such primes are now known as *Germain primes*. The first few Germain primes are 2, 3, 5, 11, 23 and 29. For example, 29 is prime and so is $(2 \times 29) + 1 = 59$. It's very natural at this point to wonder whether there are infinitely many Germain primes—but we don't (yet) know! There are good reasons to conjecture that there are, but the question currently remains unresolved.

So are the primes impossibly hard to understand?

I could keep going, to talk about even more problems about the primes that we don't yet know how to solve. The above problems are all about 'linear' patterns in the primes, but we can ask about other 'polynomial' patterns too. For example, are there infinitely many primes that are one more than a square? That is, primes such as 2, 5, 17 and 37. We don't know—yet! The question of whether there are infinitely many primes that are one *less* than a square turns out to be an exercise in secondary school algebra—changing just that one word 'less' into 'more' turns it from a problem for school students into an open research question! I'll leave the question about primes that are one less than a square as something for you to think about if you're interested.

I don't want to give the impression that mathematics consists entirely of simple problems that nobody can solve. I've described three (the Twin Primes Conjecture, the Goldbach Conjecture, and proving that there are infinitely many Germain primes), but there are also lots of problems that *have* been solved, and those have led on to new, increasingly complex problems that mathematicians are working on today.

CHAPTER 5

May 2013

Following Zhang's announcement that he could show that there are infinitely many pairs of primes that differ by at most 70 000 000, the question was then whether he and others could improve on this by lowering the number. Could Zhang's new ideas lead to a proof of the elusive Twin Primes Conjecture: could they show that there are infinitely many pairs of primes that differ by just 2?

In the past, mathematicians would work on something like this singly, or in pairs or small groups. They might hold a reading seminar, in which a small group in the same university or the same city would get together to pore over Zhang's paper, trying to understand the argument and to look for places where the details of the proof could be tightened up to improve on the bound. Or they might work in a more individual way, each becoming an expert in one or more aspects of the argument and so finding ways to improve on it. Over the subsequent months and years a steady flow of papers would be published, each incrementally improving on the last. Some people would be disappointed, as they found that their small improvement had been trumped by someone else getting in first with a larger improvement. Some improvements would be dramatic, reflecting additional new ideas from the authors, while others would be smaller, highlighting the difficulty of making any progress at all.

This is not what has happened this time. The internet has enabled a whole new way of working. Of course, it had already changed the way in which mathematicians work—no more sending postcards to overseas collaborators, but instead emails and Skype conversations. But recently there has been a more dramatic innovation: collaborating on mathematics *in public*.

For this problem, collaborative work began with a flurry of initial conversation on the blog *Secret Blogging Seminar*, after Scott Morrison put up a post 'I just can't resist: there are infinitely many pairs of primes at most 59 470 640 apart' on 30th May 2013. Morrison is a mathematician at the Australian National University, and is already something of a pioneer in the field of public collaboration in mathematics through his co-founding of the website MathOverflow. In fact MathOverflow was another home for discussion of Zhang's work. It allows some people to ask mathematical questions, and others to give answers. Its primary focus is research mathematics (rather than students asking for help with their school homework, for example), and it is managed by the users voting questions up and down so that the most interesting and most useful threads bubble to the top. It's quickly become a helpful tool for many research mathematicians: if you have a question that you think someone may have already answered, then MathOverflow is a very efficient way to obtain that answer. In the case of Zhang's work, someone asked about it on 20th May 2013, and over the subsequent two or three weeks a number of answers sought to explain the 'philosophy' behind Zhang's work, and in particular how he had succeeded when so many others had not.

There is something irresistible about the goal of reducing Zhang's bound, partly because progress is so quantifiably visible, and partly because, at least in the early stages, people were able to be the current record-holder (however briefly!), even without being a world expert in the relevant mathematics. It rapidly became clear that there were a number of points where the argument could be tightened up: Zhang had not tried to optimise every step. This is understandable, because for Zhang it was more important to publish his work rapidly and in the most readable way he could than to squeeze every last drop out of the argument—from his point of view, the precise value of the bound was not so important. At least in some cases, these aspects that were ripe for improvement were very amenable to computer calculation, and so accessible for people to get involved.

Morrison noted that a couple of others had already suggested improvements to Zhang's claimed 70 000 000, and he offered his own proposal to get an even better bound. That led to a stream of comments in response to his blog post, giving constructive feedback on his post and suggesting new potential improvements.

One of the early people to comment was Terry Tao, an Australian mathematician at the University of California, Los Angeles and a prominent member of the mathematical blogging community. Tao is an extraordinary mathematician: a winner of the Fields Medal in 2006 (one of the most prestigious awards bestowed by the mathematics community, often described as the Nobel Prize of mathematics), who works at the very highest levels in a remarkably wide range of mathematical fields. Tao was quickly grasping the details of Zhang's argument and spotting the most likely places where improvements could be made, and he was happy to share his insights on Morrison's blog. He suggested using the blog post as a 'clearing house for the "sport" of tweaking the bounds', since even making these minor improvements was a good way to understand Zhang's paper. This is a good illustration of the way in which mathematicians seek to understand each other's work, by actively testing the components of the argument to see why each part is as it is.

By 31st May 2013, Morrison, Morrison's computer and Tao were able to demonstrate that the bound 42 342 946 was achievable, again by tightening up little aspects of Zhang's argument. I'll say more about this after the next chapter (I need to introduce some important ideas first).

It wasn't really clear how much more could be gained by just these small incremental improvements, but on 31st May Tao wrote

> This project is curiously addictive; I guess because it is much easier to make progress on it than with more normal research projects :).

That addictive quality was going to prove crucial to the success of the ongoing collaborative effort.

CHAPTER 6

Making hard problems easier

Suppose for a moment that you're a research mathematician picking a problem to work on. (Perhaps you really *are* a research mathematician, in which case you should probably look elsewhere for advice on picking your next problem to tackle!) What do you do? Perhaps you have a problem that has cropped up in your previous research. After all, making progress on one problem often leads to several more questions. But perhaps you feel like working on something else. So you might pick a problem that you've read about in a recent research paper, or that you heard about over coffee at the last conference you went to, or that your collaborator told you about excitedly in an email the other day.

After a few hours or days or weeks in which you read, and think, and read more, and think more, and scribble on a notepad or blackboard, the problem starts to get under your skin. You really, really want to make some progress. Perhaps making progress on the problem would make you famous, perhaps it would just please the other two people in the world who are working on it with you. But by now the reason for thinking about the problem is that it won't leave you alone, that you *need* to know the answer, to understand what's going on.

Several months later, you start to feel that you understand the problem. Yes, it really can take that long to understand a problem. Not at a superficial level of understanding, but the proper, deep understanding where you start to see why the problem is difficult. You've tried several obvious lines of attack, which didn't work, and by exploring why they didn't work you start to appreciate the subtlety and intricacies of the problem. Far from being deterred, you are all the more intrigued.

How are you going to make progress on this problem, when it's clearly proving pretty challenging? Perhaps you need some more technical knowledge, perhaps that paper your colleague published a couple of months ago will help, or the one from the 1920s that you found in a dusty journal in the library, when you've pored over it, and understood it enough to see how you might be able to apply the technique to your new problem. But perhaps you haven't found that bit of mathematical machinery—or perhaps nobody has come up with it yet.

A good strategy can be to tackle an easier but related problem instead. This is one of the ways in which mathematicians make progress. Yes, sometimes there are spectacular leaps forwards, but there are also many smaller incremental steps, each building on earlier ones. For problems like the Twin Primes Conjecture and Goldbach's Conjecture, there are all sorts of ways to vary the problem that make it easier (not easy, mind you, just easier). Solving an easier problem might seem less glamorous, but the understanding that mathematicians gain from the solutions to these variants of a problem can lead to progress on the main problem itself. Mathematicians are forever varying the questions they're working on, modifying their aims as they gain a sense of what might be achievable.

Let me give you some examples of problems related to the Twin Primes Conjecture and Goldbach's Conjecture that mathematicians *have* managed to solve.

Almost primes

One way to modify the conjectures is to stop asking about prime numbers, since prime numbers are difficult. This has turned out to be extremely fruitful. What makes prime numbers special is that they have just two factors (themselves and 1). We could instead think about numbers with a couple of prime factors, such as $15 = 3 \times 5$ and $209 = 11 \times 19$.

For example, in 1966 Chen Jingrun (1933–1996) was able to show that there are infinitely many primes p for which $p + 2$ is either prime or a product of two primes (an *almost prime*—that's what mathematicians call a number that is either prime or a product of two primes). This was a hugely exciting result! While it's not the Twin Primes Conjecture, it's only slightly weaker. Chen did something similar for Goldbach's Conjecture too, by showing that from some point on, every even number is a sum of a prime and an almost prime. His work was a brilliant application of techniques from *sieve theory*, illustrating the power of ideas that have gone on to be a fundamental part of much of the work on these and related problems (I'll say more about sieve theory in Chapter 9).

The weak Goldbach Conjecture

It's not unusual when working on a problem to feel sure that you have a good strategy, but to find that the strategy isn't quite powerful enough to prove the result you want. This can be a good moment to see how much you *can* prove using that strategy, and perhaps to explore what the limitations are that prevent you going further.

For example, there is a very powerful approach in the area of additive number theory called the *Hardy–Littlewood circle method*. (Additive number theory simply refers to problems about adding whole numbers, often of the flavour 'which numbers can be written as sums of these other numbers?'.) Godfrey Harold Hardy (1877–1947) and John Edensor Littlewood (1885–1977), invariably known by their initials just as G.H. Hardy and J.E. Littlewood, were two of the most prominent British mathematicians of the first half of the twentieth century. Both spent much of their careers at Trinity College, Cambridge, although Hardy also spent a significant period in Oxford. Their collaboration is one of the most famous in mathematics—in 1947, the Danish mathematician Harald Bohr apparently said

> To illustrate to what extent Hardy and Littlewood in the course of the years came to be considered as the leaders of recent English mathematical research, I may report what an excellent colleague once jokingly said: 'Nowadays, there are only three really great English mathematicians: Hardy, Littlewood, and Hardy–Littlewood'.

One of Hardy and Littlewood's many contributions was the 'circle method' that they developed in the 1920s, when working on problems in additive number theory. (I'll say more about the circle method in Chapter 14.) It has turned out to be enormously powerful for a whole range of similar problems, and so it was not surprising that they and others tried to apply the strategy to the Goldbach Conjecture.

Hardy and Littlewood were able to make significant progress—but only up to a point. In order to apply their method, they needed certain detailed information about the behaviour of the primes, and they dealt with this by allowing themselves to assume an unproven conjecture (still unproven to this day!), closely related to the *Riemann Hypothesis*. The Riemann Hypothesis is a notoriously hard (and currently unsolved) problem in mathematics; proving the conjecture would tell us more about the distribution of prime numbers. These 'conditional' results of Hardy and Littlewood, depending on an unproven conjecture, were clearly not ideal, but were nonetheless something of a triumph for their circle method. Their results were two weaker variants of the Goldbach Conjecture. One was that (in a sense that they made precise) almost every even number greater than 4 is a sum of two odd primes. If there are exceptions, then there are very few of them. The other was that every sufficiently large odd number is a sum of three primes.

Just a few years after Hardy and Littlewood's work was published, the Russian mathematician Ivan Vinogradov (1891–1983) was able to adapt their approach, to develop it, to simplify it, to the point where he could drop the assumption of an unproven conjecture. In 1937, he showed unconditionally (not assuming anything) that every sufficiently large odd number is a sum of three primes.

At this point, I think I should explain two things. One is this phrase 'sufficiently large', and the other is the link with Goldbach's Conjecture.

Well, 'sufficiently large' just means that from some point on the result holds (for example, from a million onwards, or from a billion billion onwards). The idea is that this deals with the most problematic aspect, because in principle what remains is a finite check. One could, in theory, check by hand all the cases up to the point from which Vinogradov's Theorem shows the result. In practice, even with a modern computer this would have been impractical—the point from which Vinogradov's result is valid is so colossally large that it would take much too long to check it, even with a computer. But there's a strong sense amongst mathematicians that Vinogradov's Theorem was the most interesting part, because it made the rest into a finite problem.

How does this relate to Goldbach's Conjecture? If Goldbach's Conjecture is true—that is, if every even number greater than 4 is a sum of two odd primes—then it immediately follows that every odd number greater than 7 is a sum of three primes. (I'll leave you to think through why that is!) But the other direction does *not* follow: even if we knew that every odd number greater than 7 is a sum of three primes, Goldbach's Conjecture would not immediately follow. (Again, I encourage you to think about why this is so—it's a great way to get a feeling for the difference between the two statements.) It's therefore no surprise that the statement that every odd number greater than 7 is a sum of three primes is known as the *weak Goldbach Conjecture*: it's related, but is strictly weaker than Goldbach. Vinogradov didn't prove the weak Goldbach Conjecture fully (because of the 'sufficiently large' in his result), but he did get pretty close.

Since Vinogradov's paper of 1937, various mathematicians have managed to decrease the value from which it was known that the weak Goldbach Conjecture holds, but these bounds were still way beyond the reach of the most powerful computers available today. But in the last few years mathematicians have managed to close the gap, to prove the weak Goldbach Conjecture in its entirety. By the end of 2013, just a few months after Zhang's momentous announcement on bounded gaps between primes, the weak Goldbach Conjecture was finally resolved by the Peruvian mathematician Harald Helfgott (then at the École Normale Supérieure in Paris), with help from David Platt (at the Heilbronn Institute for Mathematical Research at the University of Bristol, after a slightly unusual career path that saw him studying a Mathematics degree and PhD in his 40s). The gap here is between the theoretical bound ('from this point on we know that every odd number is a sum of three primes') and what can be achieved computationally ('we've checked the result for every odd number up to this point'). Helfgott pushed the Hardy–Littlewood circle method very hard, bringing the theoretical bound within the realms of possibility, and Helfgott and Platt improved on the computational techniques in order to be able to reach Helfgott's theoretical bound. Realistically, this is a little less exciting than Zhang's work (in the sense that it was inevitable, rather than startling), but it is still a hugely impressive piece of work—and it's rather satisfying to know for sure that every odd number greater than 7 is a sum of three primes.

Amusingly, this means that we know that every whole number greater than 1 is a sum of at most four primes—this follows from the weak Goldbach result. Take your favourite whole number greater than 1. If it's odd, then either it's prime or the weak Goldbach result tells us that it's a sum of three primes and we're done. If it's even and not 2 (which is already prime) or 4 (which is a sum of two primes), then subtract 3. What's left is an odd number that is a sum of at most three primes, and so the original number is the sum of these primes plus the prime 3. This result is not as strong as the Goldbach Conjecture, but it's still pretty impressive!

The Prime Triples problem

Here's a problem for you to tackle: the Prime Triples problem.

You might have noticed that 3, 5 and 7 are all prime, and the gaps from 3 to 5 and from 5 to 7 are both just two, like two pairs of twin primes that overlap at 5. It's a threesome of prime lily pads that's very easy for a frog to jump. The Prime Triples problem asks whether there are any more of these 'prime triples', groups of three primes with gaps of two (such as 3, 5, 7). It's your turn to be a mathematician. What do you think?

I'd like to encourage you to pause reading for a few minutes, while you take some time to think about the problem. So here's a picture of a frog navigating a prime triple.

OK, ready to move on?

The Prime Triples problem is very similar to the Twin Primes Conjecture in flavour—and yet it turns out to be significantly easier to answer. (That's not to lessen your achievement if you *have* answered it, mind you, it's just that the Twin Primes Conjecture is really very hard.) I think that I'm not going to answer it right now, but I promise I'll do so soon. If that leaves you feeling dissatisfied that you don't know the answer—well, great, that's what it's like being a mathematician! There's no rush to read on, so keep thinking about it. Like many mathematicians, I often find it helpful to have a break to leave my subconscious to work on a problem before I go back to it.

We could ask a more general question about triples of evenly spaced primes. We've noted that 3, 5 and 7 are evenly spaced primes, but so are 3, 7, 11, and 11, 17, 23, and 29, 59, 89. These triples are great for a frog that's got its eye in to jump a particular distance consistently. What if the frog keeps going, are there quadruples of evenly spaced primes?

Yes—for example, we could keep going with 11, 17, 23, and get to 29. We can't then continue any further, because 35 isn't prime, although we could step back to 5, giving five evenly spaced primes. And we can't extend the 3, 7, 11 triple even to a quadruple, because 15 is not prime. But we have at least one quintuple of evenly spaced primes (5, 11, 17, 23, 29). Can we go further? Are there fifty evenly spaced primes? Or five hundred?

These turn out to be really interesting questions. We can ask a computer to help. As I write, the largest known collection of evenly spaced primes contains twenty-six numbers (but these numbers are rather large, so I'm not going to give them here). This is much larger than our quintuple above, but it's not all that impressive in the grand scheme of things (it's a long way from five hundred evenly spaced primes!). So the computer cannot resolve the question for us.

One of the constraints here is that in order for a collection of evenly spaced numbers to consist exclusively of primes, the spacing must be right—it must satisfy certain conditions in order to be feasible. This will be easier to discuss if I introduce the terminology that mathematicians use—the reason that mathematicians use technical language is not to exclude people, it's because it's much easier to talk about mathematical ideas if we're all using the same precise language. Mathematicians love precision, and everyday English is full of ambiguity and imprecision. So rather than talking about 'a collection of evenly spaced numbers', we use the phrase *arithmetic progression.* An arithmetic progression starts at some number, and then proceeds by jumps of equal size. This is quite a general concept, not one specifically about prime numbers, so perhaps I'd better switch from a frog to a grasshopper. The grasshopper begins at some starting point, and then jumps the same amount, the *common difference*, each time until it gets tired and stops (or potentially it keeps going forever, if the arithmetic progression is infinite).

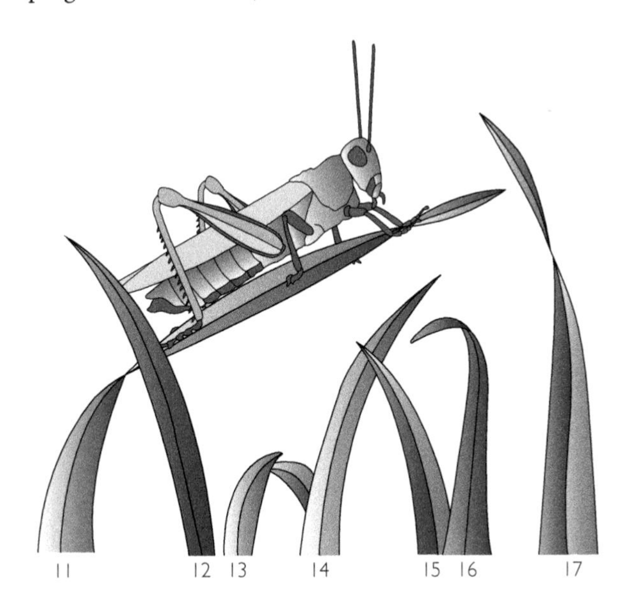

Here are some examples of arithmetic progressions.

- 1, 2, 3, 4, 5, 6, 7, 8, 9 (common difference 1).
- 5, 7, 9, 11, 13 (common difference 2).
- 11, 17, 23, 29 (common difference 6).
- −2, 0, 2, 4, 6, 8 (common difference 2).
- −17, −20, −23, −26, −29 (common difference −3).
- 22, 17, 12, 7, 2, −3 (common difference −5).

And here are some examples of sequences that *aren't* arithmetic progressions.

- 1, 2, 4, 8, 16, 32.
- 3, 5, 7, 11, 13.
- 1, 1, 2, 3, 5, 8, 13, 21.

Considering both examples and non-examples can be a very helpful way to get a feel for the scope of a concept.

I said earlier that 'in order for a collection of evenly spaced numbers to consist exclusively of primes, the spacing must be right'. In our new technical language, my assertion is that in order for every term of an arithmetic progression to be prime, the common difference must be right. It's certainly *not* the case that if I pick a clever common difference then the terms of the arithmetic progression will all be prime, but in order to stand any chance of the terms being prime we must have a suitable spacing.

Here's an example to illustrate that. Suppose I want an arithmetic progression of three primes (that is, three evenly spaced primes). Then the common difference cannot possibly be 3. Why is that? Well, in an arithmetic progression of length three with common difference 3, at least one of the three numbers is even (see Figure 6.1), and so cannot be prime unless it's 2. But then 2 would have to be the middle term of the arithmetic progression, and we find that there is no possible prime for the first term of the progression.

In fact, something more general is true: in an arithmetic progression of primes of length three or more, the common difference must be even. We can show this using a very similar sort of argument: if the common difference is odd, then at least one of the terms will be even. We can see this by checking the two possibilities for the first term.

If the first term is even, then certainly at least one of the terms is even.

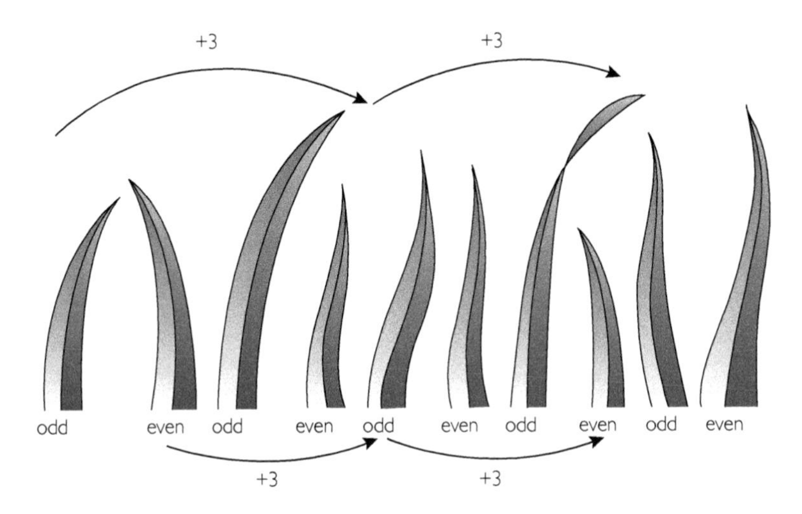

Fig. 6.1 In an arithmetic progression of length three with common difference 3, at least one of the three numbers is even.

And if the first term is odd, then the second term is the common difference added to the first term so is the sum of two odd numbers and hence even.

We can extend this idea to say more about the constraints on spacings in groups of evenly spaced primes. The argument above showing that the spacing must be even is the same sort of underlying reason behind the scarcity of prime triples (arithmetic progressions of primes with common difference 2). We know that 3, 5, 7 is a prime triple. Perhaps you convinced yourself earlier that this is the *only* prime triple. What reason did you have? There are several ways of thinking about this; I'll talk about one now and I'll mention another in Chapter 10.

In an arithmetic progression of length three with common difference 2, I claim that one of the terms must be a multiple of 3. (In fact, it's *exactly* one of the terms, but that's extra information that we don't need here.) Why is that? It's very similar to the argument with odds and evens above.

There are three possibilities for the starting number (shown in Figure 6.2): it's a multiple of 3, or one more than a multiple of 3, or two more than a multiple of 3.

We can check each case separately (see also Figure 6.3, in which each blade of grass is labelled with the remainder on division by 3, so 0 means a multiple of 3, 1 means one more than a multiple of 3, 2 means two more than a multiple of 3).

If the first number is a multiple of 3, then we're done.

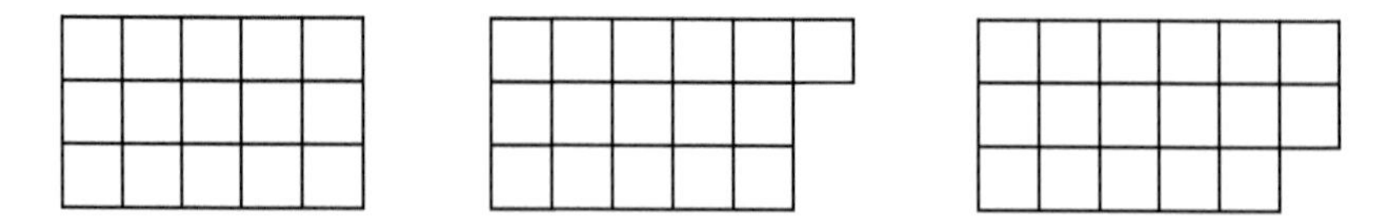

Fig. 6.2 Every number is a multiple of 3, or one more than a multiple of 3, or two more than a multiple of 3.

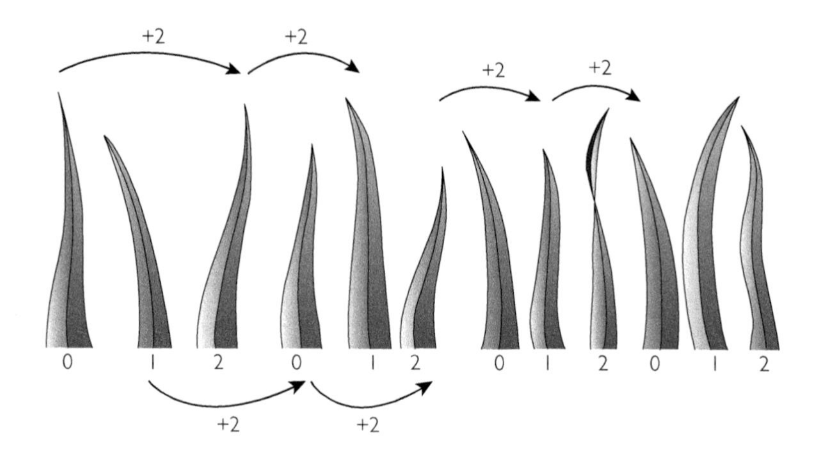

Fig. 6.3 In an arithmetic progression of length three with common difference 2, there must be a multiple of 3.

If the first number is one more than a multiple of 3, then the second term, which is two more than the first term, will be a multiple of 3.

Finally, if the first number is two more than a multiple of 3, then the third term, which is 2 + 2 = 4 more than the first term, will be a multiple of 3.

The only prime multiple of 3 is 3 itself—we saw this back in Chapter 2. So the only way to have an arithmetic progression of three primes with common difference 2 is to include 3, and by checking the possibilities we see that the only example is 3, 5, 7.

Just as we could extend our observations about arithmetic progressions with common difference 3 to a more general observation about arithmetic progression with odd common difference, we can extend our argument about progressions with common difference 2. We find that if the common difference is one or two more than a multiple of 3, then an arithmetic progression of length three (or more) must contain a multiple of 3, and so the terms can't all be prime unless one of them is 3 itself (which is quite restrictive, the first term would have to be 3).

So in general, if we go looking for an arithmetic progression of length at least three consisting entirely of primes then we should ensure that the common difference is both even and a multiple of 3. Picking a common difference that is even and a multiple of 3 doesn't guarantee that every term in our arithmetic progression is prime, but if the common difference is not both even and a multiple of 3 then we're wasting our time (there will be very few, if any, examples). To say that another way, our conclusion is that the common difference must be a multiple of 6.

For example, we saw earlier that 5, 11, 17, 23, 29 is a collection of five evenly spaced primes, with common difference 6. We could look for other common differences too. For example, 7, 19, 31, 43 gives an arithmetic progression of primes with common difference 12, and 11, 41, 71, 101, 131 is an arithmetic progression of primes with common difference 30.

This kind of reasoning can be used to analyse possible common differences for longer arithmetic progressions of primes. I encourage you to have a go for yourself—what can you say about the common difference of an arithmetic progression of five primes, for example?

Remember, though, that we're imposing conditions on common differences, but these conditions *do not* guarantee that a resulting arithmetic progression will indeed consist entirely of primes. I know that I keep saying this, but it's really important! I can take a progression with common difference 6 and first term 29 (which is prime), but even then not all the terms are prime: the next term is $35 = 5 \times 7$. So this doesn't give us a reliable way of building long arithmetic progressions of primes, but it does give us a way to narrow down the search.

This is all very interesting (at least, I think so, and I hope that you do too), but you'll notice that it doesn't address my earlier questions at all. Is there an arithmetic progression of fifty primes? We are none the wiser. We certainly haven't ruled out the possibility that there is such a progression—but we also haven't established that there actually is one. In principle, we could use a computer, but in practice we're not close to being able to do that (it would take too long), and even if we could I'd just increase the length of the target progression until it was out of reach of a computer!

So we'd like some kind of mathematical argument. Note that I've only asked 'Is there an arithmetic progression of fifty primes?'. I'd be completely satisfied with an answer 'Yes' or 'No' (with a reason, of course!). I don't necessarily need an example of a progression of fifty primes in order to answer the question. This is an important point, and one that occurs a lot in mathematics. It is often much easier to argue that some mathematical object exists than to give an explicit example of such an object.

Szemerédi's Theorem

Showing that there are arbitrarily long arithmetic progressions consisting only of prime numbers is a well-known problem, I'm going to talk a little about the history now, but if you don't want to take a little detour then you can safely skip to the next section heading! The prolific Hungarian mathematician Paul Erdős (1913–1996) conjectured that if an infinite set of positive integers has the property that the sum of their reciprocals diverges, then the set contains arbitrarily long arithmetic progressions—whatever target length we pick, we can find an arithmetic progression that is at least that long. Now, the sum of the reciprocals of the primes diverges (I'll say more about this in Chapter 10, including what I mean by a sum diverging). So the key message here is that a special case of Erdős's conjecture says that the prime numbers contain arbitrarily long arithmetic progressions. It's worth highlighting that Erdős's conjecture is essentially 'combinatorial' rather than belonging to number theory. What I mean by that is that his conjecture doesn't rely on any particular properties of the numbers in the set, but just that there are rather a lot of them (enough for the sum of their reciprocals to diverge). If Erdős's conjecture is true, then we can immediately deduce that the primes contain arbitrarily long arithmetic progressions, and this is just because there are a lot of primes, not because of deep number-theoretic properties of the primes. I find that rather intriguing!

Unfortunately, this conjecture remains unproven. A related conjecture of Erdős and Turán (going back to the 1930s) has, however, been proved, and it is now known as Szemerédi's Theorem. If we knew a sufficiently strong version of Szemerédi's Theorem, then that would give a proof of Erdős's conjecture. So I'd better tell you about Szemerédi's Theorem, and then we can get back to prime numbers.

Szemerédi's Theorem belongs to an area of mathematics called *Ramsey Theory*, named after the British mathematician Frank Ramsey (1903–1930), which is all about finding structure where there seems to be none. (Yes, Ramsey really did manage to initiate a new area of mathematics before dying so prematurely.) Ramsey Theory includes many beautiful ideas, but I'm only going to talk about Szemerédi's Theorem here, since this is what is relevant to our quest to understand prime numbers.

Suppose I take 1% of the numbers from 1 to 1 000 000 (choosing the numbers in any way I like). Am I guaranteed a 'long' arithmetic progression amongst my chosen numbers? Szemerédi's Theorem tells us that no matter how small a percentage of numbers I take, if I choose that percentage from a large enough set then I can be sure to finding a long arithmetic progression. How big 'large enough' is here depends on the percentage of numbers I choose and the length of arithmetic progression I hope to find, and this dependence is captured by the 'bounds' in Szemerédi's Theorem. If we knew enough about the bounds in Szemerédi's Theorem, and if we knew that the set of primes satisfied those conditions, then we'd be able to prove the result we want.

Terminology is only there to help us, and I'm getting slightly tired of writing 'arithmetic progression'. So let's do what the mathematicians who work on these things do, and refer to such an object by the more convenient, if less explicit, term *AP*. This still just means a sequence of evenly spaced numbers, but we're now old friends with this idea so we can afford to be a bit less formal. We can also use it to capture the idea of an arithmetic progression of a particular length—for example, we call an arithmetic progression of length 3 a *3AP* (this simply means a triple of three evenly spaced numbers).

As you might guess, Szemerédi's Theorem was first proved by Szemerédi. Endre Szemerédi is a Hungarian mathematician, who in 1975 proved the Erdős–Turán conjecture. His proof was extraordinary and ingenious, using combinatorial ideas in unexpected ways, and those ideas have led to many subsequent results proved by other mathematicians. But his proof was not quite as informative as we might hope. The theorem tells us that if we choose 1% of the numbers between 1 and N, where N is large enough, then we're guaranteed to have a 47AP, for example (and similarly for any positive percentage and any length of AP). Szemerédi was able to prove that this is true, but he was not able to say how large N needs to be—only that there is a value that is large enough.

Just two years later, in 1977, Hillel Furstenberg of the Hebrew University of Jerusalem used tools from an entirely different branch of mathematics called ergodic theory to give another proof—but this one also doesn't give us information about how large N needs to be! It wasn't until 2001 that we got these bounds, when the British mathematician Tim Gowers, of the University of Cambridge, came up with yet another way to prove Szemerédi's Theorem. For his work on this problem and others, Gowers received the Fields Medal in 1998.

Gowers's proof was exciting, both because it gave bounds and also because it gave mathematicians another way of thinking about the problem. However, as with Zhang's work on gaps between primes, the bounds were a major breakthrough but were not as good as we might hope. Since Gowers's proof, there has been a flurry of activity on Szemerédi's Theorem. There are now several proofs using very different areas of mathematics, as well as various incremental improvements on the bounds. Terry Tao (whom we met in Chapter 5) has described Szemerédi's Theorem as a Rosetta stone of mathematics: studying the similarities and connections between the various strategies of proof has led us to a much deeper understanding of the interrelationships between various parts of mathematics. But until someone manages to improve the bounds far enough, we cannot use this approach to show that there are arbitrarily long arithmetic progressions of primes.

Prime 5APs

Perhaps you've had your own personal breakthrough, thinking about the question I left you with earlier. What can you say about the common difference of an arithmetic progression of five primes? Here are my thoughts on this question.

We already know that the common difference (spacing between terms) should be a multiple of 6 for there to be a plentiful supply. What if the common difference is *equal* to 6?

By checking cases (thinking about the remainder when we divide the first term by 5), we find that in any arithmetic progression with five terms and common difference 6 there must be a multiple of 5.

Since 5 is the only prime multiple of 5, we discover that 5, 11, 17, 23, 29 is the *only* arithmetic progression with five terms and common difference 6.

The problem is that by repeatedly adding 6 we cycle round all the possible remainders on division by 5, and so after five terms we're guaranteed to have met a multiple of 5. Something very similar happens whenever the common difference is not a multiple of 5, and so either we'll find no prime 5APs or we'll find exactly one. For example, 5, 17, 29, 41, 53 is a prime 5AP with common difference 12. The only way we could have a prime 5AP with common difference 18 is if it starts with 5, so it would have to be 5, 23, 41, 59, 77—but 77 isn't prime. So there are no prime 5APs with common difference 18.

If we want to have at least two prime 5APs with a particular common difference, then we'll need the common difference to be a multiple of 6 and also a multiple of 5, so it must be a multiple of 30.

And then we can think about even longer progressions ...

More about evenly spaced primes

I started a couple of sections back by asking whether there are fifty evenly spaced primes, or five hundred. I hope that you now have a sense of the difficulty of this problem!

And yet ... we do know the answer to this one. In the first few years of the twenty-first century, the British mathematician Ben Green and Terry Tao (whom we have already met) proved a spectacular result: there are arbitrarily long arithmetic progressions of prime numbers! You pick a target, and the Green–Tao Theorem guarantees that there is an arithmetic progression of that length in which every term is prime. You want a million evenly spaced primes? No problem, the Green–Tao Theorem tells us that there is such a progression. Isn't that amazing?

So how did they do it? Well, Green and Tao were very familiar with Szemerédi's Theorem, and in particular with Gowers's work on it (indeed, Tim Gowers was Ben Green's PhD supervisor). They managed to use those ideas, and to combine them with a detailed understanding of the prime numbers in order to deduce their theorem. Their proof doesn't give examples of these long arithmetic progressions, but it tells us that if we look far enough down the list of primes then eventually we'll be able to find five hundred evenly spaced primes (or any other number of your choice). It doesn't tell us how to find those primes, but in fact having an example of five hundred evenly spaced primes is not so interesting—what's important is that we know they exist.

The Green–Tao Theorem was a major breakthrough in the study of prime numbers, and contributed to Ben Green and Terry Tao winning numerous prizes, including the Ostrowski Prize in 2005 (subsequently won by Yitang Zhang in 2013). Like all such breakthroughs it has led to a lot of follow-up work, by Green and Tao, and by many others. Mathematicians are seeking to see what the new ideas and insights can tell us about other problems, and indeed what they can tell us about prime numbers and about the original theorem of Szemerédi. Yes, understanding the primes is difficult, but every so often there are big breakthroughs, and more frequently there are little breakthroughs, small incremental steps, happening every day around the world.

CHAPTER 7

June 2013

By the end of May 2013, there had already been good progress on improving Zhang's argument in order to get much better bounds. That work concentrated on finding *admissible sets*, which play an important role in the argument, so let me tell you what they are.

Punch-cards and admissible sets

Imagine that we have all the positive whole numbers (1, 2, 3, ...) written out in a long line on a strip of paper. We're going to make a sort of stencil that we'll use to look for primes. I'm picturing something that looks like an old-fashioned punch-card, with holes cut at certain points, as illustrated in Figure 7.1.

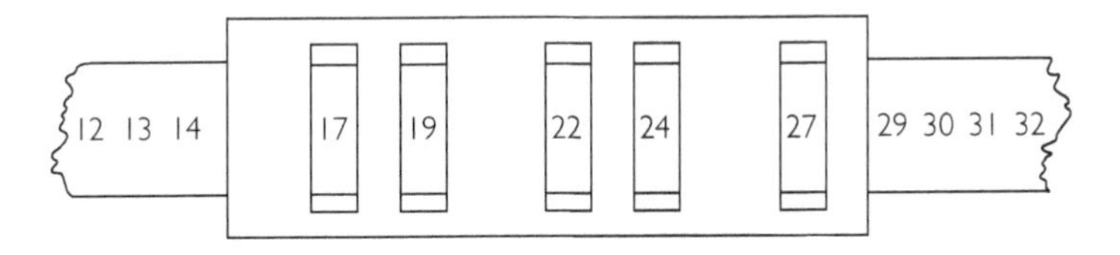

Fig. 7.1 A punch-card.

We'll slide our punch-card along the line of numbers, looking to see when all the visible numbers are prime.

For example, suppose our punch-card has three slots with gaps of 2 (see Figure 7.2).

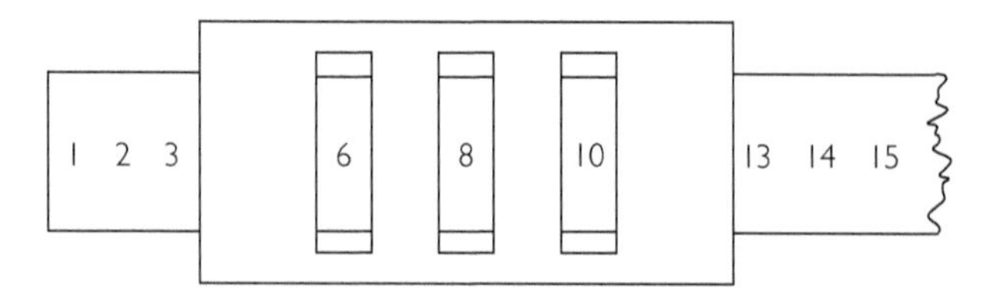

Fig. 7.2 The punch-card corresponding to $\{0, 2, 4\}$.

Then, as we saw in the last chapter, all the visible numbers are prime only when the punch-card is placed over 3, 5, 7. If we slide the punch-card to the right, one of the visible numbers will be a multiple of 3 bigger than 3, and such a number isn't prime. We represent this punch-card by the set $\{0, 2, 4\}$: the idea is that we're always looking at the numbers n, $n + 2$ and $n + 4$ for some starting number n. We represent the punch-card shown in Figure 7.1 by the set $\{0, 2, 5, 7, 10\}$.

The punch-card $\{0, 2, 4\}$ is bad from our point of view. We'd like a punch-card that gives *infinitely many* sets of visible numbers that are all prime.

Unfortunately, that's too much to hope for, because *proving* that we have such a punch-card is very difficult (there is no example with two or more holes for which we can currently prove this). But we can look instead for punch-cards where it's conceivable that there are infinitely many sets of visible numbers that are all prime—where there's no obvious reason to rule them out. Let's say such a punch-card is an *admissible* punch-card.

Our sample punch-card $\{0, 2, 4\}$ above is definitely not admissible—we have seen that there's an 'obvious' reason why there are not infinitely many sets of visible numbers that are all prime. We say that the prime 3 is a *witness* to the inadmissibility of this punch-card: one of the visible numbers must always be a multiple of 3. The problem is that the visible numbers between them always cover all possible remainders on division by 3 (those remainders are 0, 1 and 2), and so must always include a multiple of 3.

This is what I mean by an 'obvious' reason that makes a punch-card inadmissible. If there's some prime p so that the set of visible numbers always includes every possible remainder on division by p, then there must always be a visible multiple of p. So we know that there are not infinitely many sets of visible numbers that are all prime (as we saw in the previous chapter).

For example, any punch-card containing both odd and even numbers is doomed: it is inadmissible, as witnessed by the prime 2. The example illustrated in Figure 7.1, which corresponds to the set $\{0, 2, 5, 7, 10\}$, is such an inadmissible punch-card.

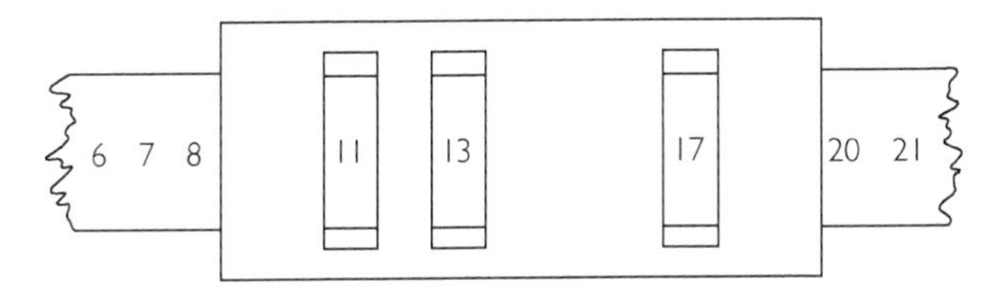

Fig. 7.3 The punch-card corresponding to $\{0, 2, 6\}$.

On the other hand, if there is no such prime p, then it's conceivable that there are infinitely many sets of visible numbers that are all prime (although we can't currently know that there really are infinitely many such sets!). And that's what *admissible* means. In the mathematical literature, authors write about *admissible sets* rather than admissible punch-cards.

For example, the punch-card $\{0, 2, 6\}$ (shown in Figure 7.3) *is* admissible. Because it contains only three numbers (three holes), we need worry only about the primes 2 and 3. For any larger prime, there are more than three possible remainders on division by that prime, and so the holes in the punch-card cannot possibly cover them all. For the prime 2, we see quickly that we're safe—either the visible numbers are all odd or they're all even, but we definitely cannot have both appearing in the same triple of visible numbers. Something similar holds for the prime 3 too: we cannot get all three possible remainders from these three holes, because the first and last numbers leave the same remainder on division by 3.

I'd like to reiterate that we don't know whether for this punch-card there are infinitely many sets of visible numbers that are all prime—if we could prove that there are, then in particular we'd have a proof of the Twin Primes Conjecture! All we're saying is that there isn't an obvious reason for there not to be infinitely many such triples, and that's what qualifies this punch-card as admissible.

Thinking about admissible sets leads to another conjecture that has not yet been proved, called the *Hardy–Littlewood Prime Tuples Conjecture*, which predicts that if we slide a punch-card corresponding to an admissible set along the strip of numbers, then there are infinitely many sets of visible numbers that are all prime. This is a generalisation of the Twin Primes Conjecture, which relates to the admissible set $\{0, 2\}$.

Goldston, Pintz and Yıldırım

The idea of using admissible sets predates Zhang. Back in 2005, a paper of Daniel Goldston, János Pintz and Cem Yıldırım made dramatic progress on the problem of understanding small gaps between primes, using the idea of admissible sets (which was around even before then). Little did anyone know it at the time, but their ideas were later to prove crucial as ingredients of Zhang's work. But their work was profound in itself, and at the time was a significant breakthrough.

A classic strategy in mathematics, when working on a seemingly intractable problem, is to assume some other unknown result. You say 'If only I knew this other statement was true, then I'd be able to prove my theorem'. Such theorems are described as *conditional*: they depend on missing information. This is a very helpful way of making progress, because it can indicate which conjectures are significant, and can help to identify where the real difficulties in an argument lie. There are many papers in mathematics that start by assuming the Riemann Hypothesis (one of the most famous unproven conjectures in mathematics), for example. There are some that start by assuming that the Riemann Hypothesis is false. There are even some rare unconditional theorems that we know are true, where there are two arguments. One proceeds by assuming that the Riemann Hypothesis is true, and the other by assuming that it is false. One of these must be a valid proof, and so the theorem is true!

The Riemann Hypothesis, or strictly speaking the Generalised Riemann Hypothesis, is potentially useful for the Twin Primes Conjecture too. In the 1980s the British mathematician Roger Heath-Brown proved that if Siegel zeros exist, then there are infinitely many pairs of twin primes. (Don't worry about what Siegel zeros are, it doesn't matter right now!) Now, the Generalised Riemann Hypothesis predicts that these Siegel zeros don't exist, but we don't yet have a proof. So Heath-Brown's paper demonstrates that it's sufficient to concentrate on showing that if there are no Siegel zeros then the Twin Primes Conjecture is true. This is not how the recent work has been proceeding, but it's a nice link with the Riemann Hypothesis, which I'll mention again in Chapter 8.

The conjecture that Goldston, Pintz and Yıldırım relied on is the Elliott–Halberstam Conjecture. I'll say a little more about this in Chapter 10. For now, it's enough to know that it is a detailed prediction about the distribution of the primes. Assuming a suitable form of the Elliott–Halberstam Conjecture, Goldston, Pintz and Yıldırım were able to prove that given a sufficiently large admissible set (one that contains enough elements), there are infinitely many sets of visible numbers that contain at least two primes.

It's hard to appreciate the importance of this, especially in the light of the more recent work by Zhang. At the time, it was the first paper showing bounded gaps between primes, even assuming an unproven conjecture, and so was a major breakthrough. Assuming the strongest form of the Elliott–Halberstam Conjecture, they were able to show that there are infinitely many pairs of primes that differ by at most 16. Just 16—wow! This bound is much better than Zhang's 70 000 000, but what makes Zhang's bound special is that it doesn't depend on any unproven conjectures.

Goldston, Pintz and Yıldırım proved a number of results in their paper (assuming Elliott–Halberstam), including the bound of 16 as well as other theorems. But arguably one of the most important aspects of their paper was not the results they proved, but the journey they took along the way. They demonstrated a significant link between bounded gaps between primes and the distribution of the primes as predicted in the Elliott–Halberstam Conjecture. That was spectacular—and paved the way for future discoveries.

Improving on Zhang's bound

Zhang's work built on that of Goldston, Pintz and Yıldırım. His key step was to reduce the assumption needed: instead of using the full power of the Elliott–Halberstam Conjecture, Zhang showed that the result would follow from a weaker statement about the distribution of the primes—and then crucially he was able to prove this weaker statement. What was most startling about this was that everybody 'knew' that this approach was not viable! Most experts in the field had ruled this out as a strategy, and so overlooked the idea that turned out to be fundamental to Zhang's argument. Anyway, more about this later—let's go back to admissible sets.

As I described in Chapter 5, others (including Morrison and Tao) had managed to improve on Zhang's bounds. Their insights were all about finding suitable admissible sets. Zhang showed that if an admissible set is large enough, then there are infinitely many sets of visible numbers that contain at least two primes. If, for example, we have a punch-card that has width 1 000 000 and that has enough holes (as specified by Zhang's theorem), then we can slide the punch-card along to different positions, and infinitely many times we'll find a visible set that contains at least two primes, so we can deduce that there are infinitely many pairs of primes that differ by at most 1 000 000. So the punch-card must contain enough holes (the admissible set must be large enough)—but at the same time we want to make it quite narrow, because the width of the punch-card dictates the bound obtained.

As an illustration, here is an admissible set with width 558 and 100 holes. This is much too small for Zhang's theorem to apply, but has the advantage of being small enough that I can include it here—I wouldn't be able to reproduce an example that is large enough to be covered by Zhang's result! Even this is too large for me to sketch it as a punch-card.

$$\{0, 6, 12, 22, 28, 40, 42, 46, 48, 52, 60, 66, 70, 82, 88, 90, 96, 106, 108, 118,$$
$$120, 126, 130, 132, 136, 138, 148, 160, 162, 166, 172, 178, 186, 192, 196, 208,$$
$$210, 216, 220, 222, 228, 238, 246, 250, 252, 262, 270, 280, 286, 288, 292, 298,$$
$$306, 312, 318, 328, 330, 342, 346, 348, 358, 360, 370, 372, 376, 382, 390, 396,$$
$$400, 402, 406, 412, 420, 426, 430, 438, 442, 448, 456, 460, 462, 468, 472, 480,$$
$$496, 502, 508, 510, 516, 522, 526, 528, 532, 538, 540, 546, 550, 552, 556, 558\}$$

There are thus two natural ways to try to improve on Zhang's bound. One is to decrease the size of admissible set required (the number of holes in the punch-card). The other is to find a narrower admissible set with enough holes.

The early improvements on Zhang's bound all came by exploring possible admissible sets and finding narrower large sets. One convenient way to create a large admissible set is to choose a bunch of suitably large prime numbers and use those to create the punch-card: $\{p_1, p_2, \ldots, p_k\}$. This is guaranteed to lead to an admissible set, and so the work comes in measuring the width of such a set of given size. This was the approach that Zhang took, which led him to his bound of 70 000 000 (although it was soon pointed out that a closer analysis of his admissible set gives a better bound). Australian number theorist Timothy Trudgian, and then Morrison, Tao and others (building on a paper of Douglas Hensley and Ian Richards dating right back to 1973) were more careful about the construction of narrow punch-cards with many holes. Here, a computer comes in very handy to check particular examples: for once this is a problem where there are few enough things that need checking that a computer can take on the tedious work! (Checking the admissibility of a particular punch-card took a few hours of computer time, or sometimes less when it became clear that the punch-card would not work.)

But maybe punch-cards with fewer holes could also be acceptable? Improving on that aspect of Zhang's work would require a detailed understanding of his argument.

Polymath

Is massively collaborative mathematics possible? This was the question posed by the mathematician Tim Gowers in a blog post at the end of January 2009. Like Terry Tao, Gowers writes a mathematical blog that is followed by a large number of people, including many professional mathematicians. We met Gowers briefly in the last chapter—he gave a new and important proof of Szemerédi's Theorem. This particular blog post, suggesting a new way for mathematicians to collaborate on a problem, attracted a great deal of attention from those within the mathematics community and also from others interested in similar ideas in other disciplines. Indeed, Gowers refers to a blog post by Michael Nielsen, a writer and scientist who discussed open and large collaborations in contexts beyond mathematics. Gowers had been thinking about the idea for some time, and Nielsen's blog post and the increasing interest in related ways of working (such as the Open Notebook Science movement) spurred him to action. Gowers went further than speculating about what might be possible. He described the type of collaboration that he envisaged, which he called Polymath, outlined the ground rules, and went on to suggest a mathematical problem for the Polymath project to tackle. The rules themselves give an insight into both how mathematicians do mathematics, and how Gowers imagined the Polymath project working.

His vision was that the process of solving a hard mathematical problem could be subdivided into solving many smaller problems, which could potentially be solved by many people together, with no one person having to work too hard and with no one person overseeing work, a bit like an ants' nest. Potentially participants would not need to have an overview of the whole project, but rather could specialise in their own little sub-problems. One of his rules said

> The ideal outcome would be a solution of the problem with no single individual having to think all that hard. The hard thought would be done by a sort of super-mathematician whose brain is distributed amongst bits of the brains of lots of interlinked people. So try to resist the temptation to go away and think about something and come back with carefully polished thoughts: just give quick reactions to what you read and hope that the conversation will develop in good directions.

Gowers was very keen to encourage participants to share their immediate thoughts, rather than working on ideas independently in private, with an emphasis on expressing their immediate thoughts as clearly as possible so that others could build on them. He wrote

> When you do research, you are more likely to succeed if you try out lots of stupid ideas. Similarly, stupid comments are welcome here. (In the sense in which I am using 'stupid', it means something completely different from 'unintelligent'. It just means not fully thought through.)

One of the striking innovations of Polymath is this sharing in public of ideas that have not yet been carefully considered or checked. Normally, mathematicians publish only their successful and polished ideas: the others end up in the recycling bin, or erased from the board. With Polymath, the whole point is to share all those ideas in public. Maybe another participant can quickly see why an idea is not going to work, and therefore that line of attack can be closed off, or perhaps one participant can see how to develop another's idea into something useful.

Collaborations sometimes come unstuck because participants disagree about how credit for the work should be assigned. Gowers addressed this from the outset:

> Suppose the experiment actually results in something publishable. Even if only a very small number of people contribute the lion's share of the ideas, the paper will still be submitted under a collective pseudonym with a link to the entire online discussion.

Some disciplines, such as high energy physics, are used to producing papers with tens or even hundreds of authors: papers from the Large Hadron Collider at CERN routinely have huge numbers of authors, for example. This has not been the case in mathematics (particularly pure mathematics), where many papers have a single author, and a paper with five or six authors is unusual.

Polymath is pioneering a new way of working, and this presents challenges to the mathematical community. For example, when someone applies for an academic job, they list their publications as part of their CV. Referees might comment on the merits of these papers, and on the extent to which the applicant contributed in the case of joint papers. How would Polymath papers be assessed? In one way it would be easier to evaluate the level of contribution—if an appointments committee wanted to know the extent of an applicant's contribution to a particular piece of work, they would no longer have to rely on the description provided by the applicant or by a referee, because they could simply check the online discussion and see *exactly* what the applicant contributed.

Doing mathematics is not about doing long division with bigger and bigger numbers, it's not about plugging in some numbers and running through a routine calculation. The process is both messy and creative—mathematicians need to have ideas, perhaps many ideas, when working on a problem. They then have to check patiently to see whether any of the ideas work (perhaps after adaptation). Sometimes it takes a long time to discover that an idea doesn't work. In practice, this means that if several mathematicians work on the same problem they may well individually go through the same process of establishing that a particular idea doesn't work. Ideas that don't work don't get published, so there's no record of failed attempts for future mathematicians. The great mathematician Gauss is quoted as saying

> no self-respecting architect leaves the scaffolding in place after completing the building.

Having struggled through failed attempts and faced dead ends on the way to a solution to a problem, the mathematician erases the unsuccessful parts from the whiteboard (or tosses them in the recycling bin) and writes up only the successes for publication. Consequently, there is no mechanism for mathematicians to record work along the lines of 'We tried this approach and it doesn't work because of the following obstruction', and so future generations of mathematicians may well find themselves unknowingly retracing paths that lead nowhere. While this may seem inefficient, the process of trying an idea and understanding why it doesn't help can actually be very useful when working on a problem, because it can give greater insight into the mathematical situation and may lead to further ideas that might be more fruitful. One of the revolutionary features of Polymath is that all the dead ends and diversions along the way are recorded for posterity, in complete detail, for others to see.

In subsequent blog posts, Gowers went on to set out the mathematical background to an unsolved problem, to describe that problem (giving a combinatorial proof of the Density Hales–Jewett Theorem), and to describe an idea that he had for solving it. The problem was a significant one—experts in the area were genuinely interested in solving it—but it wasn't the most famous one around (not nearly as famous as the Twin Primes Conjecture, for example). It's a fascinating area of mathematics, but I'm not going to allow myself to take a detour to describe it here, since the problem itself is not related to prime numbers. In fact, it relates to Ramsey Theory and to Szemerédi's Theorem, which we saw briefly in the last chapter.

What *is* relevant about the Density Hales–Jewett Theorem is that there is now a combinatorial proof. (The theorem had been proved previously by Hillel Furstenberg and Yitzhak Katznelson using ideas from ergodic theory; what was important here was that the new proof took a very different route.) To the surprise of many, including Gowers himself, the Polymath project came up with a proof, within just a few weeks. Gowers wrote on 10th March 2009 that 'I am basically sure that the problem is solved'. An experienced mathematician can know that a problem is solved even before every t is crossed and every i dotted. Sometimes you just *know* that it'll all work out, and that what remains is checking and straightening out of details.

Through a large number of comments on blog posts, and eventually also a wiki set up to keep track of progress, the participants managed to solve a serious research problem that had remained unsolved for years. The paper was then written up under the pseudonym D.H.J. Polymath (DHJ for Density Hales–Jewett), and was published in the *Annals of Mathematics* (the same journal that was later to publish Zhang's work on twin primes). Research problems get solved all the time in mathematics. Uniquely, this time anyone in the world can see all the work on the problem, not just the final paper.

More Polymath projects

After the unexpected success of the Polymath project on the density Hales–Jewett problem, it was inevitable that other projects would follow, as mathematicians tried to learn from DHJ Polymath and to see what types of problem might be amenable to this sort of massive collaboration.

The next few Polymath projects were on problems from various parts of mathematics, with varying degrees of success. I don't want to give a blow-by-blow account of these, and anyway the nature of Polymath projects is that you can find the full details online! I do want to highlight two projects, though, because I think that they're particularly informative.

One is a type of 'mini-polymath' project that Terry Tao started. Every summer several hundred school-age mathematicians from around the world gather to compete in the International Mathematical Olympiad (IMO). The event lasts several days, with many opportunities for getting to know teams from other countries, but the academic focus is on two extremely difficult exams, each lasting four-and-a-half hours and each containing just three problems.

In July 2009, Tao suggested taking the final (and traditionally most challenging) problem, Question 6, from that year's IMO as the basis for a Polymath-type discussion, because 'the problem is receptive to the incremental, one-trivial-observation-at-a-time polymath approach'. An IMO problem is somewhat different from a research project, because there were already plenty of people who knew a solution to the problem (and it was known that it could be solved using the kinds of mathematical techniques that IMO contestants know, and plausibly could be solved by a rather exceptional contestant within the time allowed for an IMO paper). There was consequently an emphasis on not spoiling the problem by looking up (or thinking of) a solution and then posting it, but otherwise it was structured much like the first Polymath project. Tao gave some ground rules, in addition to Gowers's original Polymath rules, most of which revolved around not looking up solutions, not giving away solutions, not linking to solutions, and so on, but he also described what kinds of posts *would* be welcome:

> if you have a potentially useful observation, one should share it with the other collaborators here, rather than develop it further in private, unless it is 'obvious' how to carry the observation further.
>
> Actually, even 'frivolous' observations can (and should) be posted on this thread, if there is even a small chance that some other participant may be able to find it helpful for solving the problem.
>
> Similarly, 'failed' attempts at a solution are also worth posting; another participant may be able to salvage the argument, or else the failure can be used as a data point to eliminate some approaches to the problem, and to isolate more promising ones.

There are not many contexts in which mathematicians are asked to share 'failed' attempts at a solution, but are explicitly requested *not* to share a complete solution!

Solving an IMO problem is a real challenge, but inevitably not the same sort of challenge as solving a previously unsolved research problem, and so within a few days the discussion had wrapped up having found multiple solutions. This accelerated process makes it a little easier to reflect on what works and what does not work in a collaborative project like this, and Tao wrote a post considering these matters. Some of his comments were around technical matters: a linear blog conversation is sometimes restrictive for a mathematical conversation that is trying to go in several directions at once, and having a wiki or similar where good ideas and key points can be recorded is immensely helpful. (The participants in the first Polymath project found much the same thing.)

Thinking about differences between Polymath and more traditional approaches, Tao wrote

> Polymath projects tend to generate multiple solutions to a problem, rather than a single solution. A single researcher will tend to focus on only one idea at a time, and is thus generally led to just a single solution (if that idea ends up being successful); but a polymath project is more capable of pursuing several independent lines of attack simultaneously, and so often when the breakthrough comes, one gets multiple solutions as a result.

He went on to add

> Polymath progress is both very fast and very slow.

Polymath ideas come thick and fast, but it can be hard for participants to absorb and process suggestions from others, and so sometimes promising directions are lost, or temporarily overlooked, in the sea of ideas. An individual who hits on a promising direction might chase it through to a successful conclusion more quickly than Polymath can untangle the threads, but Polymath is better at generating lots of avenues to explore, and at spotting mistakes and dead ends, than one person working alone.

Polymath5

The other Polymath project I'd like to highlight is Polymath5, on the Erdős discrepancy problem (named after Paul Erdős, whom we met in Chapter 6). I'm not going to go into the mathematics here, because that would be a diversion from our main story. However, I'd like to tell you about the project at a high level, because I think that it's illustrative of ways in which mathematics makes progress, and possibly helps to bring into focus the question of what really counts as progress.

At the end of December 2009, Gowers outlined three potential new Polymath projects on his blog, and invited readers to vote. The result of this was that in early January 2010 work began in earnest on the Erdős discrepancy problem, and this became Polymath5. One attractive feature of this problem for Polymath was that there is a computational aspect: those with the relevant skills could seek large examples using computers, and these were extremely helpful for the project. In mid-January, when discussion was flowing fast, Gowers used his experience of Polymath5 to observe that

> ... the process is ideal for problems where there is the potential for an interplay between theory and experiment. When looking at DHJ [the first Polymath project], we looked at both theory and experiment, but the two were fairly disjoint. This was due to the nature of the problem: it is not computationally feasible to gather data about DHJ except in very low dimensions, and it is therefore difficult to draw any general conclusions from the results. But for the Erdős discrepancy problem the situation is quite different, and we have learned a huge amount from looking at long sequences that have been generated experimentally.

In fact, this was even before the project had 'officially' started!

Gowers formally started the project with a post summarising the conversations to that point, which had included a large experimental component that could be used to inform (and be informed by) the future theoretical work. These kinds of summary blog posts, capturing the key ideas from the comments on the previous posts, have proved to be enormously helpful to the success of Polymath projects, whether written by Gowers or Tao or someone else—whoever was hosting the project. It's hard to see how they can be avoided, although they are not entirely in the spirit of the ants' nest with no overall director. Anyway, conversations continued apace on Polymath5. All sorts of interesting sub-problems emerged along the way, as so often happens in mathematics, and some (but not all) were rapidly solved. By the middle of February, after various ebbs and flows, conversation slowed somewhat, but there were still exciting new ideas that gave participants something to think about. The project still continued to explore these ideas over the following months, although it had lost some of the initial frenzy and excitement of the early weeks of the project.

In June 2010, Gowers revitalised the conversation by posting on his blog to share some thoughts he'd had on the problem. It highlights a question that individual researchers ask themselves frequently: how do you decide when to stop working on a problem? You're not deciding to give up, just to set one project aside for a while and to work on something else instead. Famously, our subconscious is an effective mathematician. The great mathematician Henri Poincaré (1854–1912) had been working on a problem for some time when the solution came to him just as he was climbing onto a bus:

> At the moment when I put my foot on the step the idea came to me, without anything in my former thoughts seeming to have paved the way for it.

Setting a research project aside for a while and then revisiting it is a good strategy, but how do you decide when to do it? And in the brave new world of Polymath, how do the participants agree when to stop a project, or at least to suspend work on it? This has a particular importance for Polymath because it seems probable that mathematicians are unlikely to work on a problem by themselves (away from Polymath) while it is the subject of an active Polymath project, so it may be necessary to reach some sort of closure, even if the project might subsequently re-awaken. It might also be helpful to write up the work of the project for publication, not least so that if it leads into a future piece of work then it can be referenced carefully in the conventional way.

By early September 2010, Gowers was asking other Polymath participants how to proceed (close the project, or continue, or some sort of option between those extremes), because while he was still interested, there were fewer active participants in the project as a whole. Polymath projects seem to begin with a very intense period, in which ideas come rapidly and participants have to devote a lot of time simply to keep up with the discussion. But very few people are able to maintain that level of involvement, simply because they have other demands on their time. By this stage the Erdős discrepancy project had rather fizzled out, although plenty of people were still aware of it and were no doubt thinking about it in private from time to time. There was a mini-revival in the summer of 2012, and occasional updates along the way, but the Polymath project went rather quiet.

And then in September 2015 Terry Tao announced that he had a solution of the Erdős discrepancy problem! I attribute this to a coming together of three key factors: the work of Polymath5; an exciting new piece of work by two early-career mathematicians, Kaisa Matomäki and Maksym Radziwiłł; and great insights and ideas by Tao himself, building on a blog comment by the German mathematician Uwe Stroinski. Tao had been a very active participant in the Polymath5 project, and was consequently deeply immersed in the ideas that had been discussed there. This kind of immersion is vital when working on a problem, and in this case it meant that Tao was well placed to make the breakthrough a few years later.

Matomäki and Radziwiłł were not working on this problem at all. Their spectacular breakthrough was on a seemingly entirely separate problem in number theory. Following her PhD at Royal Holloway in the UK, Matomäki has returned to a research fellowship in Finland; Radziwiłł did his PhD at Stanford in the US and since then has had prestigious academic positions in the US and Canada. As a result of their highly successful collaborative work, they jointly received the 2016 SASTRA Ramanujan Prize 'given annually to a mathematician not exceeding the age of 32 for outstanding contributions in an area of mathematics influenced by the late Indian mathematical genius Srinivasa Ramanujan'. (We've met other winners of this prize in this book: Terry Tao won in 2006 and Ben Green in 2007.)

Having learned of their work, Tao worked with them on various applications of their new results to other problems, and in early September 2015 he wrote a blog post about a paper that he had just written jointly with Matomäki and Radziwiłł, in the same way that he blogs about each of his papers. On 9th September 2015, Uwe Stroinski (who had himself participated in Polymath5) commented on this blog post, observing that

> The Sudoku-flavor arguments remind me on [sic] the EDP Polymath project,

and going on to ask whether the work by Matomäki and Radziwiłł might help with the Erdős discrepancy problem. A few hours later, Tao replied explaining why Matomäki–Radziwiłł would not apply. Roll on another few hours, and Tao had changed his mind:

> Hmm, actually there does appear to be a connection between the EDP and something we looked at in our previous paper.

By 11th September 2015, Tao had written up his ideas into a new blog post, drawing on the work of Polymath5 to show that a certain unproven conjecture, which he called the 'non-asymptotic Elliott conjecture', would give an answer to the Erdős discrepancy problem. A week later, on 18th September, Tao put up another blog post, about two papers he had just posted online. The first proved a result related to (but slightly weaker than) the non-asymptotic Elliott conjecture, building on his joint work with Matomäki and Radziwiłł. The second showed that this theorem answers the Erdős discrepancy problem. Success!

Tao is an extraordinary mathematician. His capacity to absorb deep and technical new ideas, and then to apply them rapidly to difficult problems is truly remarkable. But the above story neatly illustrates a question about how one gives credit in mathematics. Tao will forever be known as the person who solved the Erdős discrepancy problem (amongst his many other accomplishments), and of course he did. But this time it is perhaps even more evident than usual just what the ingredients going into the solution were, because of the public discussion via Polymath5 and then on Tao's blog. I have no wish to lessen his achievement, and it's important to stress that Tao was scrupulous about acknowledging the work of Polymath5 and the suggestion from Stroinski in his paper. But I do think it's important that as a mathematical community we recognise the many ways in which people contribute to solving a problem. Some, like Tao, are brilliant at having ideas and flashes of insight that finish off solutions, others make connections and recognise similarities (as Stroinski did here), others carry out experimental work on computers, explore the patterns and structures revealed within the resulting data, and make bold conjectures (as many Polymath5 participants did), and still others prove results that turn out to have unexpected consequences (as Matomäki and Radziwiłł did). It's tempting to be seduced by the romantic stories of lone figures making solo breakthroughs—to give all the credit to the person who put the final piece in the jigsaw—but that is to overlook the preliminary work on which those mathematicians built. There is no glamour in sorting out the edge pieces of a jigsaw, but it can be an enormously helpful step in completing the puzzle.

So, was Polymath5 a success? The project did not solve the Erdős discrepancy problem. But it did generate a lot of useful data and ideas, and meant that several people, including Stroinski and Tao, were immersed in the problem and so were able to seize the moment when it came, when new results and techniques had been developed by others. By my reckoning, this undoubtedly qualifies as a success. Gowers has said that without Polymath5 he thinks we'd still be waiting for a solution to the Erdős discrepancy problem. This is important, because it means that understanding a piece of mathematics better is a legitimate aim for future Polymath projects (as well as for future projects of individual mathematicians, of course). A project on a famous unsolved problem might seem pointless if the experts in the area all agree that the problem is out of reach with our current knowledge, but if the aim is not to solve the problem but rather to produce a really detailed description of our current understanding of the problem, something that future researchers can use to kickstart their own thinking, then the aim is both achievable and worthwhile.

Polymath8

I'd like to tell you about another Polymath project, this time proposed by Terry Tao. Given his enthusiasm for Polymath, and his expertise in the study of prime numbers, it was no surprise when he proposed Polymath8, which he called *Bounded gaps between primes*. His initial proposal from 4th June 2013 suggested two (intertwined) goals for the project:

1. Further improving the numerical upper bound on gaps between primes; and
2. Understanding and clarifying Zhang's argument (and other related literature, e.g. the work of Bombieri, Fouvry, Friedlander, and Iwaniec on variants of the Elliott–Halberstam conjecture).

Let me expand on that briefly. Goal 1 was to improve on Zhang's bound of 70 000 000 (always remembering that the dream target is just 2). It's illuminating that Tao explicitly mentioned a second goal, of making sense of Zhang's work and the many other papers on which he draws. Mathematics in textbooks is (hopefully) clear and easy to follow, but mathematics does not arrive in that neatly polished form. When someone first proves a theorem, it can take months or years or decades for other mathematicians to make sense of the ideas. This is not about checking whether a proof works (although that can sometimes take ages too), rather it is about *understanding* an argument—about seeing where the ideas come from, how they link with previously understood ideas, what the right definitions are, where there are clever shortcuts in proofs, and whether there are alternative statements of results that become easier to prove. All this takes real effort on the part of mathematicians, and Tao was proposing that this could also be part of the collaborative Polymath work.

Within just hours of Tao making the proposal, mathematicians from around the world started getting stuck in—albeit with a short diversion to discuss how the Polymath project might sit alongside more conventional attacks on the problem being made by others. The Polymath model is so new in mathematics that mathematicians are still thinking about how the practical aspects work. For example, what if a Polymath project ends up proving something that an individual mathematician was on the point of publishing?

And so work on Polymath8, called *Bounded gaps between primes*, began in earnest. Zhang started the ball rolling in May 2013 by showing that there are infinitely many pairs of primes that differ by at most 70 000 000. By the end of May this bound had come down to 42 342 946. Could the Polymath participants bring that down still further, maybe even as far as 2 (for the Twin Primes Conjecture)?

In early June 2013, as the project got started, Polymath participants created a 'league table' on a wiki. It starts by recording Zhang's paper proving that there are infinitely many pairs of primes that differ by at most 70 000 000, recorded with the date and Zhang's name and the bound 70 000 000. Each time a Polymath participant thought that they had an improved bound, they filled in a new row with their claimed figure, and then other Polymathematicians (to coin a word) checked the argument, and either confirmed it or expressed their doubts. The league table is online via the wiki for the project 'Bounded gaps between primes' (I've given the web address in the Further Reading at the end of the book), and you'll see that there are quite a few numbers crossed out as people retracted their claimed bounds or amended them following feedback from others. Mathematicians don't always get it right first time, and part of the point of Polymath is to share ideas quickly before they're all fully checked so as to allow others to see them—even though this is quite uncomfortable behaviour for many mathematicians who are used to working very hard to make only true statements in public!

The numbers in the league table started falling dramatically in early June 2013, as mathematicians started to understand the details of Zhang's argument and so identified aspects that could be improved. Zhang's paper contained a chain of reasoning involving a whole string of quantities, each of which depended on one or more previous quantities in delicate ways. One of these quantities was called k_0: Zhang was able to show that given an admissible set of size k_0, there are infinitely many sets of visible numbers that contain at least two primes. So one goal was to find a very narrow punch-card with at least k_0 holes, and this is how the initial improvements were made (by finding narrower punch-cards with enough holes). The next natural goal was to try to shrink k_0: if punch-cards with fewer holes will do, then we can again seek very narrow punch-cards with enough holes and this will improve the bound still further. One of Zhang's key contributions was to prove a sort of weakened form of the Elliott–Halberstam Conjecture for a particular parameter, which he was able to take to be 1/1168. So one approach was to improve on this result (by increasing the parameter to something larger than 1/1168), but another was to improve the dependence: taking the result for 1/1168 as given, is it possible to chase through the dependencies of parameters on each other to decrease the acceptable value of k_0? In the first few days of June, this is exactly what happened—and by 6th June 2013 the bound was down to 387 620. Wow!

But then things took a different turn. Mathematicians are human, and so sometimes mathematicians make mistakes. On 6th June, Hungarian mathematician János Pintz (of Goldston–Pintz–Yıldırım fame) put a 'preprint' on the arXiv. Papers in mathematics, as in other disciplines, are submitted to journals and peer-reviewed before being accepted for publication, but it is hugely convenient to share pre-publication drafts before this process is finished (it is not uncommon for the review process to take several months), in order to share the work with others. The arXiv is a website designed for this purpose: authors upload their draft papers, which are carefully timestamped; authors can also upload new versions as necessary, but the old versions are preserved in the archive. Pintz's preprint offered new improvements on Zhang's work—not unexpectedly, since he is one of the experts in this kind of mathematics—and others excitedly seized on this to try to make even more progress with the Polymath project.

Alas, there were problems with the draft, and so a whole swathe of proposed bounds had to be retracted as the work of Pintz on which they were based was found to contain errors. It's hard to convey just how fast the conversation was moving: mathematicians were being swept along in the excitement, and rushing to claim ideas and try to get a new world record. Polymath was ideal here, because it meant that people were sharing their ideas in public and checking each other's work, in a much more efficient way than is usually the case.

The Polymathematicians soon bounced back, though, and in fact Pintz's ideas still turned out to be useful to the project. By the end of June 2013 the figure was down to just 12 006, which is a massive improvement on Zhang's bound of 70 000 000 in a very short space of time, although still quite some way off the Twin Primes Conjecture's goal of 2. Would Polymath be able to get it even lower?

CHAPTER 8

How many primes are there?

How many primes are there? This is a silly question, because we saw back in Chapter 2 that there are infinitely many primes. We even saw Euclid's proof of this theorem, so there's no doubt about it. But a good way of getting some insight into the distribution of the primes is to ask how many primes there are *up to some point*. How many primes are there up to 100? Up to 1000? Up to 1 000 000? Up to x? This was a question that preoccupied mathematicians during the late eighteenth and nineteenth centuries, and it was tackled by some of the great names of the age: Adrien-Marie Legendre (1752–1833), Carl Friedrich Gauss (1777–1855), Peter Gustav Lejeune Dirichlet (1805–1859), Pafnuty Chebyshev (1821–1894) and Bernhard Riemann (1826–1866), to name but a few. Work on the problem continues to this day.

I'm not going to go into the full story here, because that would be a book in itself (and there are several good books out there that discuss it), but it is relevant for our quest to understand gaps between prime numbers, so I'll give edited highlights.

One of the triumphs of nineteenth century mathematics came right at the end of the century, in 1896, when the Belgian mathematician Charles de la Vallée Poussin (1866–1962) and the French mathematician Jacques Hadamard (1865–1963) independently proved the Prime Number Theorem. Mathematicians do sometimes solve famous old problems! I'll state the theorem now, and then explain it afterwards.

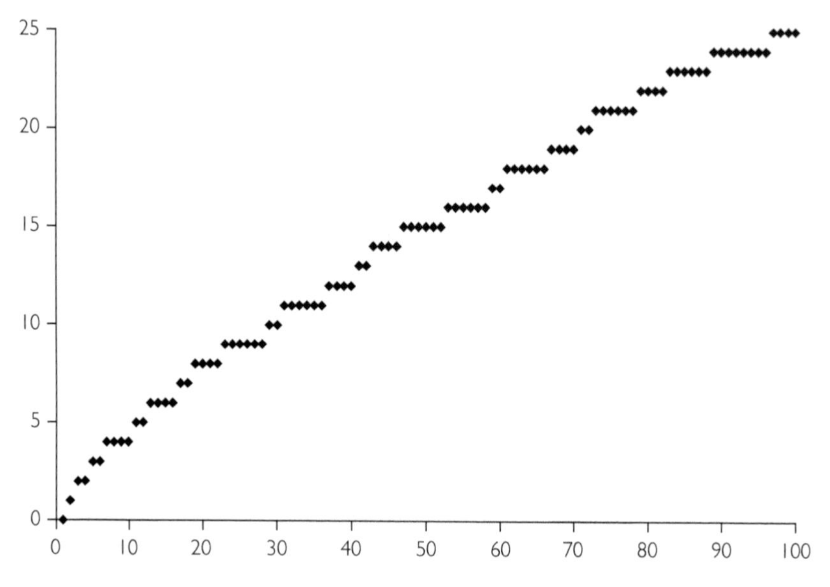

Fig. 8.1 The number of primes less than or equal to x for each value of x up to 100.

Theorem (The Prime Number Theorem) *We have*

$$\pi(x) \sim \frac{x}{\log x}.$$

What does that mean? The informal version, which is all we need for this story, is that when x is very large (that is, *huge*), the number of primes up to x is approximately some function of x that we can easily work out, namely $\frac{x}{\log x}$. Figure 8.1 shows the number of primes less than or equal to each value up to 100 (note the sudden jump at each prime—it's not at all a smooth function), and Figure 8.2 shows the function $\frac{x}{\log x}$ for x up to 100 000 to give a sense of the shape of the curve.

This is great if you know what $\log x$ means, and not so helpful otherwise. The logarithm is a standard function (you'll find it on your pocket calculator), and it's an increasing function—as x gets bigger, so does $\log x$ (although $\log x$ increases rather slowly). If you don't already know about logarithms, don't worry about it: we don't need to go into the detail here. If you do already know about logarithms, then I should explain that I've written $\log x$ for the natural logarithm of x, the logarithm to the base e. This is also sometimes written as $\ln x$, but number theorists traditionally write $\log x$ so that's what I'm doing.

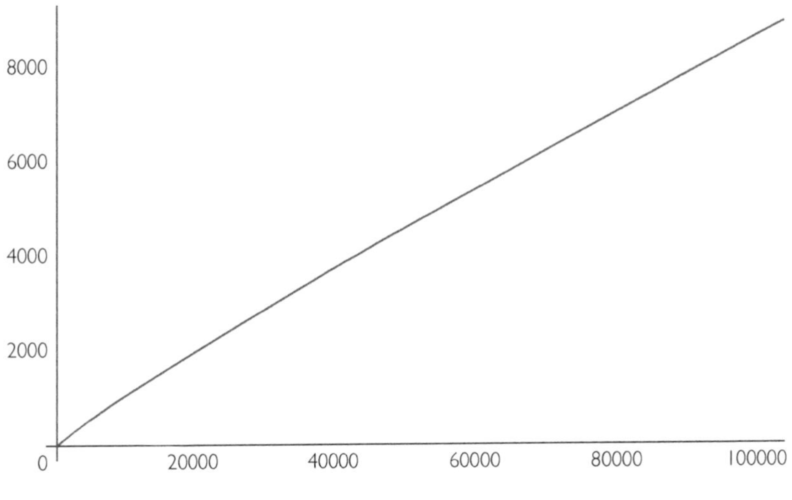

Fig. 8.2 $\frac{x}{\log x}$ for x up to 100 000.

The standard notation for the number of primes up to x is $\pi(x)$. So the number of primes up to 100 is $\pi(100) = 25$, for example. This π has nothing to do with the π we're familiar with from the formula for the area of a circle, it's just a convenient letter that someone decided to use for this function a long time ago and that stuck. It could have been $f(x)$ or $p(x)$ or all sorts of other things, but convention now dictates that it's $\pi(x)$.

In case you're interested, here's a more careful explanation of the statement of the Prime Number Theorem. The $\sim$ symbol means that the two functions are asymptotically equal—when the variable x is very large, the left-hand side is approximately the right-hand side, and the approximation gets better and better as x gets larger. Formally, it means that the left-hand side divided by the right-hand side tends to 1 as x tends to infinity.

By the way, you should not use this formula to estimate the number of primes up to a small number, it's helpful only for large values of x. If you want to know the number of primes up to a small number, work them out and count them—or, more sensibly, ask your computer to work them out and count them for you!

This approximation for the number of primes up to a given point had been conjectured for some time, and various people had managed to make partial progress towards proving it. We might wonder how it is that two people (Hadamard and de la Vallée Poussin) independently proved the same result at the same time, using the same sort of approach. The answer here is that the key tools that would lead to their proofs had become available over the preceding few decades, and they were the first two who managed to push through the details to use those tools to prove the theorem. Their arguments both rely on ideas from an area of mathematics called *complex analysis*. Crudely, this is calculus with complex numbers—I say crudely because in fact complex analysis is very beautiful, full of astonishing theorems that ought not to be true but somehow are thanks to the subtlety of differentiation of complex functions and the behaviour of the logarithm in the complex plane. (When I was an undergraduate, I remember reading about the complex analysis course that I was about to study, and the description said that the subject was very beautiful. I felt a bit short-changed during the course, because I found it very hard and not at all beautiful. But now that I've thought about the material a lot more, I think I see why that course description said what it did. I have similar feelings about some pieces of music that I've played: they did nothing for me at first, but as I immersed myself in the music I felt I understood it better and it really grew on me.)

Now, where was I? Ah yes, complex analysis. Halfway through the nineteenth century, Riemann and others demonstrated how the new subject of complex analysis could be invaluable for deepening our understanding of prime numbers, not least via a function now called the Riemann zeta function. I'm not going to make any effort to describe the Riemann zeta function here, we don't need to get into the technicalities. It was by proving certain detailed properties of the Riemann zeta function, and in particular where it takes (or rather doesn't take) the value 0, that Hadamard and de la Vallée Poussin were able to prove the Prime Number Theorem.

One of the most famous unsolved problems in the whole of mathematics is the Riemann Hypothesis, which makes a very specific prediction about where the Riemann zeta function takes the value 0. If we could prove the Riemann Hypothesis, then we'd be able to update the Prime Number Theorem with a much more precise version, giving an even better approximation for the number of primes up to a particular value.

One consequence of the Prime Number Theorem is that on average the gap between two consecutive primes of size around x is approximately $\log x$. The 'on average' is extremely important here. We've seen that sometimes consecutive prime numbers are very close together, and indeed to a large extent this book is all about that phenomenon. It's also the case that sometimes consecutive prime numbers are a long way apart. (Here's a problem for you to think about. Can you show that there are one hundred consecutive numbers that are *not* prime? That shows that there's a pair of consecutive primes that are at least 100 apart. I'll answer this question, and return to other questions about primes being far apart, in Chapter 16.) But the Prime Number Theorem says that *on average* the gap between the nth prime and the next prime is about the logarithm of the nth prime. In particular, this is an increasing function, so the average gap between consecutive primes grows: on average, the primes become more spread out. That certainly matches with my intuition that it's harder for larger numbers to be prime.

How many twin primes?

It was silly to ask how many primes there are, because we know that there are infinitely many, but we made it a sensible question by asking how many primes there are up to any given number. The conjecture is that there are infinitely many pairs of twin primes. Can we say (or predict) anything about how many pairs of twin primes there are up to any given number?

It turns out that we can make a very good prediction of the number of pairs of twin primes up to any given number, but it's just a *heuristic*. By modelling the primes, we can make a prediction, and the numerical data strongly suggests that it's a good prediction, but we can't (yet!) prove that it's a correct estimate.

The idea is to model the primes as though they are random. Of course, the primes are *not* random: any given number is either prime or is not prime, we don't get to toss a coin to decide which. But this model, known as *Cramér's model* after the Swedish mathematician Harald Cramér (1893–1985), is helpful for getting a feel for what's going on. I'll try to describe this in some detail, because I think it's interesting to see this way of thinking about the primes, but you can skip straight to the punchline at the end of this section if you don't want to think about the reasoning at this point!

The Prime Number Theorem tells us that for numbers around x (thinking of x as large), the probability that a given number is prime is about $\frac{1}{\log x}$. So for a number n around x, the probability that n is prime is about $\frac{1}{\log x}$, and the probability that $n + 2$ is prime is about $\frac{1}{\log x}$. If these events were independent, we'd see that the probability that n, $n + 2$ is a pair of twin primes is about

$$\frac{1}{\log x} \times \frac{1}{\log x} = \frac{1}{(\log x)^2},$$

and so the number of pairs of twin primes up to x is about $\frac{x}{(\log x)^2}$.

Unfortunately, this is nonsense!

The problem is that exactly the same argument would predict that the number of pairs of primes n, $n + 1$ up to x is about $\frac{x}{(\log x)^2}$. But we know exactly how many pairs of primes n, $n + 1$ there are up to x, namely just one. The pair 2, 3 has the right form, but any pair of larger consecutive numbers contains an even number that is necessarily not prime.

So there is something not right about this argument.

The problem is that the events we're considering are not independent. If n is prime, then $n + 2$ is almost certainly odd, and this affects the probability that it is prime.

If we choose any arbitrary two numbers, then the probability that each is odd is $\frac{1}{2}$, so the probability that both are odd is $\frac{1}{2} \times \frac{1}{2} = \frac{1}{4}$.

However, we are choosing two numbers n and $n + 2$ that are separated by 2, so the probability that they're both odd is $\frac{1}{2}$ (because $n + 2$ is odd if and only if n is odd).

We consequently need to multiply our estimated probability by a correction factor of

$$\frac{\frac{1}{2}}{\frac{1}{4}} = 2.$$

But a recurring theme in this book is that it's not enough to think about odd and even, we need to think about divisibility by other numbers too. For example, if n is one more than a multiple of 3, then $n + 2$ is almost certainly not prime (because it's a multiple of 3).

Let's use the same sort of argument as above, but this time our aim is to find numbers that are not divisible by 3, rather than odd numbers (numbers that are not divisible by 2). The probability that a number is divisible by 3 is $\frac{1}{3}$, so the probability that a number is *not* divisible by 3 is $1 - \frac{1}{3}$.

If we choose any arbitrary two numbers, then the probability that each is not divisible by 3 is $1 - \frac{1}{3}$, so the probability that they are both not divisible by 3 is $(1 - \frac{1}{3})^2$.

But we're choosing two numbers n and $n + 2$ that are separated by 2. In order for them both not to be divisible by 3, we need to choose n so that neither n nor $n + 2$ is divisible by 3. So the probability that n and $n + 2$ are not divisible by 3 is $1 - \frac{2}{3}$.

So this time we need to correct our estimated probability by multiplying by

$$\frac{1-\frac{2}{3}}{(1-\frac{1}{3})^2} = \frac{\frac{1}{3}}{\frac{4}{9}} = \frac{3}{4}.$$

We can extend this argument to think about divisibility by an arbitrary small prime p (with $p > 2$). If we choose any two numbers, then the probability that they are both not divisible by p is $(1-\frac{1}{p})^2$, but we're choosing two numbers that are separated by 2, so the probability that these two numbers are not divisible by p is $1-\frac{2}{p}$. We consequently need to correct by a factor of

$$\frac{1-\frac{2}{p}}{(1-\frac{1}{p})^2} = \frac{\frac{p-2}{p}}{\frac{(p-1)^2}{p^2}} = \frac{p(p-2)}{(p-1)^2}$$

for each odd prime p, and by a factor of 2 for the prime 2.

So our prediction for the number of pairs of twin primes is our previous prediction of $\frac{x}{(\log x)^2}$ (where we assumed that n and $n+2$ behaved independently) multiplied by a correction factor for each prime. This leads to the following prediction for the number of pairs of twin primes up to x: it should be approximately

$$C\frac{x}{(\log x)^2},$$

where C is the so-called *twin prime constant*

$$C = 2 \prod_{\substack{p \text{ prime} \\ p \geq 3}} \frac{p(p-2)}{(p-1)^2}.$$

The $\prod$ symbol on the right means to take a product: for each prime $p \geq 3$, we multiply by the appropriate factor. It's just a shorthand notation for the idea we developed above. I'd read the expression for C out loud as '2 times the product over all primes p greater than or equal to 3 of p times $(p-2)$ divided by $(p-1)^2$'. The product contains infinitely many terms (one for each odd prime), but the larger primes make a very small contribution, and the product converges. (I'll say more about convergence in the context of sums in Chapter 10. You can safely not worry about it for this product—if you weren't losing sleep over the matter, then please continue to sleep soundly!)

When we do the calculation (with the help of a computer), we find that the twin prime constant C is approximately 1.320 .... We can then gather data about the actual number of pairs of twin primes and compare it with this prediction—and the prediction turns out to be extremely good indeed.

There is another way to generate the same prediction, which is always an encouraging sign. This alternative approach proceeds using the Hardy–Littlewood circle method, about which I'll say much more in Chapter 14. In principle, this argument could give a proof, not just a prediction. The problem is that we cannot get good enough estimates using that approach for this particular problem, so for now it remains a prediction.

This is, however, all rather compelling evidence (but definitely not a proof!) for the Twin Primes Conjecture, because the predicted number of pairs of twin primes up to x is $C\frac{x}{(\log x)^2}$, which grows pretty fast as x grows, and in particular tends to infinity. If only we could prove it! Let's check back in on progress towards finding a proof of the Twin Primes Conjecture.

CHAPTER 9

July 2013

One of the very attractive properties of this work on gaps between primes is that it's extremely easy to track the progress being made. All too often in mathematics the only visible sign of progress on a problem is the publication of a paper after the work has been done. It's just the nature of the subject—if you've spent two years trying to prove a theorem and haven't yet proved it, then there may well be nothing to report in public. It may even be hard to identify progress to yourself in private! The work of Polymath is all done in public, so in some sense it's possible to see what's happening, but it's still often difficult to know how far off a solution might be (certainly for those not working on a problem, but sometimes also for those doing the work). With the Polymath project on bounded gaps between primes, it's easy to measure progress. The project started at 70 000 000 (the bound that Zhang obtained), with the aim of reaching 2. By the end of June 2013, Polymath had got down to 12 006.

A recurring feature of Polymath is that the most successful projects have had a computational component, allowing people with different sorts of expertise to get involved. The skill of writing some efficient code to do a lengthy calculation in a reasonable amount of time is a valuable one, and a number of participants have been happy to help in this area even though they were not experts on the main mathematical problem.

As I described earlier, the early progress came from finding narrower admissible punch-cards with enough holes. By this stage, Polymathematicians had started making real inroads on the problem of decreasing the number of holes required. Every time someone decreased the required number of holes,

k_0, in the online league table, others could deploy their computers to help find the narrowest possible admissible set with the newly revised number of holes, leading to a corresponding new world record on gaps between primes.

By the end of July 2013, the acceptable value of k_0 had shrunk from Zhang's original 3 500 000 to just 632, and the resulting very narrow admissible sets meant that the project had proved that there are infinitely many pairs of primes that differ by at most 4680. That's a massive improvement in just a couple of months, and a real testament to the power of Polymath to harness minds in a very effective collaboration.

The parity problem

I have a sort of confession to make. Polymath wasn't going to reach 2 (that is, the Twin Primes Conjecture) just by tweaking Zhang's work.

When you're stuck on a problem, when you have a technique or strategy in mind but it's just not working out, a really productive thing to do is to think hard about *why* that technique or strategy is not working. Why is this productive? You develop a deeper understanding of a strategy if you understand its limits as well as its successes, if you can get a clearer picture of problems that it can solve and also of problems that it can't.

Zhang's work relies fundamentally on a bunch of ideas known as *sieve theory*—the work of Goldston, Pintz and Yıldırım on which Zhang built also uses sieve theory, and in fact sieve theory is now the standard approach for problems in this area. (I must stress, however, that thinking 'Oh, I know, I'll use sieve theory' is easy; picking a suitable sieve and getting it to work requires profound understanding, insight, creativity and technical excellence!) One of the pioneers of sieve theory was the Norwegian mathematician Atle Selberg (1917–2007), who in the 1940s found new and hugely effective ways to apply the idea of a sieve.

This links back to the Prime Number Theorem, which we met in the last chapter (it's the one that gives an approximate formula for the number of primes up to any given value). The proof of the Prime Number Theorem in 1896 by Hadamard and de la Vallée Poussin relied on complex analysis, which in some ways felt rather unsatisfactory. Do we really need tools from calculus and complex numbers in order to count prime numbers? And so mathematicians began the search for a so-called 'elementary' proof of the Prime Number Theorem. This is 'elementary' in a technical mathematical sense meaning 'not using complex analysis', rather than the sense in which Sherlock Holmes said (or didn't say) 'Elementary, my dear Watson'.

I think of the resolution of this search as something of an anticlimax, to be honest, but that's not the fault of the people who found the answer! In 1948, Selberg used his sieve methods to prove a certain asymptotic formula relating to the distribution of the primes. Within a few months, Selberg and Erdős had gone on to give an elementary proof of the Prime Number Theorem. Unfortunately, this led to a dispute over who had priority: each drew on ideas from the other, but they disagreed over how the work should be published (as a joint paper or separately), perhaps at least in part because Erdős was by disposition a prolific collaborator, whereas Selberg preferred to work by himself.

So there is an 'elementary' proof of the Prime Number Theorem, and it is definitely of interest and uses some ideas that mathematicians have been able to apply to other problems. But it is very far from easy, and when university lecturers teach the Prime Number Theorem to undergraduates (usually undergraduates towards the end of their degree) they tend to use the approach using complex analysis and the Riemann zeta function.

Nonetheless, Selberg's work has gone on to be hugely influential in number theory, and sieve theory is now a key part of the toolkit of analytic number theorists. Selberg went on to identify a limit of his approach, called the *parity problem*.

The *parity* of a whole number is simply whether it is odd or even. For example, it is very convenient to talk about two numbers that have the same parity, meaning that they are both odd or both even.

The parity that Selberg considered was the parity of the number of prime factors of a given number. For example, a prime number always has exactly one prime factor, so has an odd number of prime factors, whereas a number such as 15 that is a product of two primes has an even number of prime factors. What Selberg discovered was that the sieve techniques he was championing were unable to distinguish effectively between numbers with an odd number of prime factors and numbers with an even number of prime factors, and this is what is known as the parity problem. For the project to show bounded gaps between primes, it seems that existing sieve theory methods alone will not get below 6: it's plausible (but highly optimistic) that they will be able to show that there are infinitely many pairs of primes that differ by at most 6, but reaching the full Twin Primes Conjecture is going to need some new ideas.

Still, by the end of July 2013 Polymath, building on all the previous work, had managed to show that there are infinitely many pairs of primes that differ by at most 4680. How much lower could they go?

CHAPTER 10

What's so mathematical about my mathematical pencil?

Deep and fascinating mathematics can arise from simple beginnings.

A few years ago, someone gave me a rather intriguing pencil—with no explanation of its mathematical significance.

The pencil has a hexagonal cross section: it has six sides, each with a row of numbers on it, as I've sketched in Figure 10.1.

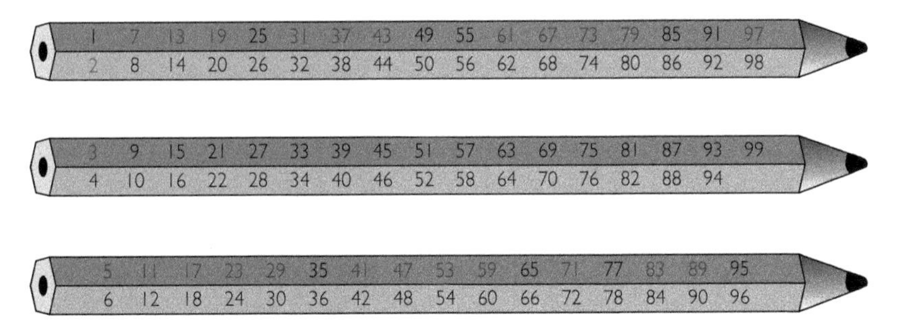

Fig. 10.1 My mathematical pencil.

Having looked at the pencil for a moment, I saw that the numbers run 1, 2, 3, ..., spiralling round the pencil so that 7 is next to 1, and 8 is next to 2, and so on. Some of the numbers are red while others are black, and having peered carefully I recognised the red numbers as being our old friends the prime numbers. In case the sketches of the pencil aren't clear enough, let me type up the numbers as they appear on the pencil (Figure 10.2).

1	7	13	19	25	31	37	43	49	55	61	67	73	79	85	91
2	8	14	20	26	32	38	44	50	56	62	68	74	80	86	92
3	9	15	21	27	33	39	45	51	57	63	69	75	81	87	93
4	10	16	22	28	34	40	46	52	58	64	70	76	82	88	94
5	11	17	23	29	35	41	47	53	59	65	71	77	83	89	95
6	12	18	24	30	36	42	48	54	60	66	72	78	84	90	96

Fig. 10.2 The numbers on my mathematical pencil.

This is reminiscent of the grid we saw in Chapter 2, but there we looked at numbers in a familiar arrangement of ten columns whereas here we have six rows. As usual, as a mathematician I am interested not so much in the information displayed on the pencil, but in what it can tell me about the distribution of the primes as a whole. What would happen if my pencil was hugely longer, what would I see then?

There are various patterns we might notice here. I'll refer to the rows (the sides of the pencil) by their smallest numbers. So one pattern is that the 4 and 6 sides of the pencil don't contain any prime numbers. Would that continue if the pencil were longer?

Well, yes, because all the numbers on those sides of the pencil are even. We discussed in Chapter 2 that the only even prime is 2. So we can be confident that there are no primes on the 4 and 6 sides of the pencil, no matter how far we extend the pencil, and moreover we see that 2 is a rather sad and lonely prime, being the only prime on its side of the pencil.

Building on that, we might speculate about the prime 3, which is languishing in solitary splendour on its side of the pencil. Are there any more primes on that side if we extend the pencil?

Closer inspection reveals that every number on the 3 side of the pencil is a multiple of 3. The bottom row, the 6 row, consists of the even multiples of 3 (that is, the multiples of 6), and the 3 row consists of the odd multiples of 3.

But now we know what to do. The only prime multiple of 3 is 3 itself. So 3 indeed is alone in being prime on that side of the pencil, no matter how long the pencil is.

And now we can draw a rather remarkable conclusion. What this pencil tells us, using the above arguments, is that apart from 2 and 3, *every prime is one more or one less than a multiple of* 6. Isn't that neat? Let me say it again: apart from 2 and 3, every prime (in the whole world, not just on my rather short pencil) is one more or one less than a multiple of 6.

This is a very cute fact, and once we have thought of it it's not too hard to check that it's true (in fact, we've already justified it). But it does give us some insight into the distribution of the primes. After all, if we hadn't thought about it then we might have assumed that the primes would be evenly scattered round the pencil, rather than clustered on just two sides (plus the exceptional 2 and 3). Also, it does have some nice consequences.

Back in Chapter 6, I posed a problem about prime triples. Let me remind you. I defined a *prime triple* to be a group of three primes with gaps of 2, such as the three primes 3, 5 and 7. The problem was to determine whether there are any more prime triples. We saw later in the chapter that this is the *only* prime triple, by considering the various possibilities for the remainder when we divide the first element of a hypothetical triple by 3. Thanks to the mathematical pencil, we now have another way to see that the only prime triple is 3, 5, 7. We know that apart from 2 and 3, every prime is one more or one less than a multiple of 6. That immediately tells us that the only way we can have a prime triple is if it contains 3, and then we find ourselves back with our old friend the prime triple 3, 5, 7.

Having realised what the pencil was telling me, I might have packed it away in my pencil case and moved on to something else. But, as happens all the time in mathematics, answering one question prompted several more. Asking good questions is a really important part of mathematics, albeit one that often doesn't appear at school where the emphasis tends to be more on students *answering* questions. What kinds of questions might we ask inspired by the mathematical pencil? Here are the questions that occurred to me; perhaps you have further questions to add to this list.

- We know (from Chapter 2) that there are infinitely many primes to be split between the two categories (one more than a multiple of 6, and one less than a multiple of 6). Are there infinitely many primes of both types, or are there infinitely many of one type and only finitely many of the other? (There can't be finitely many of both types, or we'd have finitely many primes in total!)
- If you were a prime number, would you rather be one more than a multiple of 6 or one less? To put that in a more careful way, of the primes up to a million, or a billion, do we expect more that are one

more than a multiple of 6, or more that are one less than a multiple of 6, or about the same of each?

- What would happen if my pencil had 7 sides? Or 10 sides? Or 99 sides? Would we see the same sorts of things as with a pencil with 6 sides?
- For a pencil with six sides, we've seen that a side can contain no primes (as with the 4 and 6 sides), or exactly one prime (as with the 2 and 3 sides), or many (infinitely many?) primes (as with the 1 and 5 sides). Is it ever possible, by picking a pencil with a suitable number of sides, to have exactly two primes on one side of the pencil? Is there an easy way to predict which sides have no primes, which have one, which have two, which have infinitely many?

These all turn out to be questions that lead to rather interesting mathematics. Let's have a go at answering some of them.

Are there infinitely many primes that are one less than a multiple of 6? Are there infinitely many primes that are one more than a multiple of 6?

The good news is that it turns out that we can make a lot of progress on the first question using ideas that we've already discussed in this book. There *are* infinitely many primes that are one less than a multiple of 6, and we can prove this by using a slightly adapted version of Euclid's proof that there are infinitely many primes (this was the proof that we saw in Chapter 2). You might remember that the idea was to suppose that there are only finitely many primes, then multiply them all together and add 1 to obtain a number that simultaneously must have a prime factor but is not divisible by any of the primes, thereby reaching a contradiction. This time, we can do something similar, but concentrating only on primes that are one less than a multiple of 6. We suppose there are only finitely many primes of this type, multiply them all together *and multiply the result by* 6, then *subtract* 1. It is then possible to argue that the resulting number must have a prime factor that is one less than a multiple of 6, but at the same time it is not divisible by any of the primes on our original list, and so again we get a contradiction. I'll leave you to fill in the details!

Depending on how you look at it, the bad news, or the even more interesting news, is that this argument does not transfer across in the same way to show that there are infinitely many primes that are one *more* than a multiple of 6.

Why is this anything other than bad news? Well, it shows that there's some subtlety here, some difference between being one more than a multiple of 6 and one less, and so there is a piece of mathematics to unpick and to understand. As a mathematician, I think that's definitely a good thing! Understanding *why* an argument doesn't work can be a really good way to gain insight into a mathematical idea.

Happily, this one is not too difficult to unpick. The crucial point that makes the Euclid-style argument work for the 'one less' case is that any number that is one less than a multiple of 6 must be divisible by a prime number of that form, because if we multiply primes of other forms (2, 3, or one more than a multiple of 6) then we never get a result that is one less than a multiple of 6. But it *is* possible for a number that is one more than a multiple of 6 to have no prime factor that is one more than a multiple of 6, and so the argument falls apart. For example, 55 is one more than a multiple of 6, but $55 = 5 \times 11$ so its prime factors are both one less than a multiple of 6.

So is it true that there are infinitely many primes that are one more than a multiple of 6? Our first strategy for proving it has failed, but that doesn't mean the result is wrong. In fact it's true, but proving it requires some extra ideas beyond just a minor adaptation of Euclid's argument: it needs a bit more number theory. I'll leave that one for you to ponder (it's quite a challenge).

Different sorts of pencil

What happens if the pencil has a different number of sides? For the six-sided pencil, the 2, 3, 4 and 6 sides of the pencil each have at most one prime. How can we generalise that? The key point about 2, 3, 4 and 6 for the six-sided pencil is that each of those numbers shares a common factor bigger than 1 with 6 (the number of sides), and that common factor will divide every number on that side of the pencil, which makes it hard for them to be prime (the only way such a number can be prime is if it's equal to the common factor and that happens to be prime). More generally, if the row heading and the number of sides share a common factor bigger than 1, then there's at most one prime on that side of the pencil. For example, with a 15-sided pencil we'd get exactly one prime on the 3 and 5 sides, and no primes at all on the 6, 9, 10, 12 and 15 sides.

The only sides where we can hope to find more primes are those where the row heading and number of sides have no common factor bigger than 1—we say that the row heading and number of sides are *coprime*. For our hypothetical 15-sided pencil, the sides still in play are the 1, 2, 4, 7, 8, 11, 13 and 14 sides, since each of these numbers is coprime to 15.

But do those sides actually have infinitely many primes, or not? In the case of the pencil with six sides, we've seen that those sides (the 1 and 5 sides) *do* contain infinitely many primes, but it took some work to prove that (and I skipped one of the two arguments). It turns out that a more general result is true—but significantly harder to prove. Roughly speaking, Dirichlet's Theorem, from 1837, tells us that if there's no 'obvious' reason for there not to be infinitely many primes on a side of the pencil, then in fact there *are* infinitely many primes there. When I say 'obvious reason', I mean that the row heading and the pencil share a common factor, as we saw earlier. The theorem tells us that if this isn't the case, then there are infinitely many primes on the side of the pencil. This is kind of astonishing. We rule out a bunch of cases immediately (the ones where there's a common factor greater than 1), having thought of an obvious reason for them not to work. What Dirichlet's Theorem tells us is that this is the *only* reason for cases not to work—there isn't some other less obvious obstruction that we haven't thought of. That's pretty surprising, and really rather lovely.

In more formal language, we might phrase the theorem as follows.

Theorem (Dirichlet) *If a and d are coprime (have highest common factor 1), then $a + kd$ is prime for infinitely many positive whole numbers k.*

Returning to our hypothetical 15-sided pencil, the theorem tells us that each of the 1, 2, 4, 7, 8, 11, 13 and 14 sides contains infinitely many primes. Wow!

As we have already seen, we can prove some special cases of this one at a time. Proving the full theorem needs quite a lot more work, and I'm not going to talk you through that—it requires more mathematical technology than I want to introduce here. But I think that I can give you a flavour of the kind of argument that gets used.

Let's go back to the primes, rather than subsets of the primes. We've already seen Euclid's proof that there are infinitely many primes (suppose not, multiply them all together and add one, get a contradiction). As I described above, we can adapt that idea for some individual cases of Dirichlet's Theorem, but each needs its own modification, and so to prove the general result we need a new idea. I'd like to outline another proof that there are infinitely many primes, since this can be adapted more conveniently for Dirichlet's Theorem. This argument is attributed to Euler, the 18th century Swiss mathematician who's already appeared in this book as the recipient of Goldbach's letter containing his famous conjecture—it's not often that Euler gets a mention only as having received a letter, since he was one of the most prolific mathematicians in history!

Here's Euler's idea. Take all the primes in the world, take their reciprocals (that is, take 1 divided by each of them), and add up the answers. So we add $\frac{1}{2}$, $\frac{1}{3}$, $\frac{1}{5}$, $\frac{1}{7}$, $\frac{1}{11}$, and so on. We might write the sum as

$$\frac{1}{2}+\frac{1}{3}+\frac{1}{5}+\frac{1}{7}+\frac{1}{11}+\cdots=\sum_{p \text{ prime}} \frac{1}{p}.$$

Don't worry if you're not familiar with the notation on the right-hand side of the equation, it's just a shorthand for the sum on the left of the equals sign. We read the right-hand side as 'the sum over all prime p of 1 over p'.

What can we say about this sum?

To get a feel for the question, let's think about some other similar sums. The sum of the reciprocals of the natural numbers (that is, take every positive integer, do 1 divided by it, and add up the answers), is called the *harmonic series*:

$$\frac{1}{1}+\frac{1}{2}+\frac{1}{3}+\frac{1}{4}+\frac{1}{5}+\cdots=\sum_{n=1}^{\infty} \frac{1}{n}.$$

Famously, this sum *diverges*: we add up infinitely many numbers and do not get a finite answer. To say that another way, given any number, we can add up the first terms of the sequence (up to some point) to get an answer larger than the given number. We can see this quite directly, by grouping the terms:

$$\underbrace{\frac{1}{1}}+\underbrace{\frac{1}{2}}+\underbrace{\frac{1}{3}+\frac{1}{4}}+\underbrace{\frac{1}{5}+\frac{1}{6}+\frac{1}{7}+\frac{1}{8}}+\underbrace{\frac{1}{9}+\frac{1}{10}+\cdots+\frac{1}{16}}+\cdots.$$

The sum of each bundled group of fractions is greater than or equal to $\frac{1}{2}$. We can see that immediately, we don't have to do anything annoying like add up fractions. For example,

$$\frac{1}{3}+\frac{1}{4} \geq \frac{1}{4}+\frac{1}{4}=2 \times \frac{1}{4}=\frac{1}{2},$$

and

$$\frac{1}{5}+\frac{1}{6}+\frac{1}{7}+\frac{1}{8} \geq \frac{1}{8}+\frac{1}{8}+\frac{1}{8}+\frac{1}{8}=4 \times \frac{1}{8}=\frac{1}{2},$$

and similarly for the others.

And so we see that by adding enough terms from the series we can get a value larger than any number we think of—the series diverges.

This argument is absolutely classic in the area of mathematics called *analysis*, which studies concepts like limits and infinite sums (and calculus) rigorously. The flavour of analysis is all about inequalities; to show this sum diverged we used inequalities. In a moment, we'll again use inequalities, this time to show that a certain sum converges.

By way of contrast, let's think about the same sort of sum but where we only include the square numbers:

$$\frac{1}{1} + \frac{1}{4} + \frac{1}{9} + \frac{1}{16} + \frac{1}{25} + \cdots = \sum_{n=1}^{\infty} \frac{1}{n^2}.$$

Remarkably, it turns out that this sum *converges*. Even though we're adding infinitely many things, they're so small (and get even smaller so fast) that we end up with a finite number. Amusingly, it turns out that the sum is actually equal to $\frac{\pi^2}{6}$, but that detail is incidental here. All that matters for our purposes is that the sum converges. It's a bit fiddly to try to work out the value of the sum directly, but we can show that it converges by comparing it with another sum that's easier to work with.

The idea is that instead we study the sum

$$1 + \frac{1}{2} + \frac{1}{6} + \frac{1}{12} + \frac{1}{20} + \cdots = 1 + \sum_{n=2}^{\infty} \frac{1}{n(n-1)}.$$

There are two good things about this sum. One is that we can show that it converges (in fact we can compute its value), as I'll show you in a moment, and the other is that it's bigger than the sum we're interested in.

Why is it bigger? Well, each term of the new sum is at least as big as the corresponding term of the sum of the reciprocals of the squares. We can check this for the first few terms:

$$\begin{aligned} 1 &\geq 1, \\ \frac{1}{2 \times 1} &\geq \frac{1}{2 \times 2}, \\ \frac{1}{3 \times 2} &\geq \frac{1}{3 \times 3}, \\ \frac{1}{4 \times 3} &\geq \frac{1}{4 \times 4}, \\ \frac{1}{5 \times 4} &\geq \frac{1}{5 \times 5}. \end{aligned}$$

More generally, we can express this algebraically: we have $\frac{1}{n(n-1)} \geq \frac{1}{n^2}$ for each $n \geq 2$. Thus if the new sum converges then so must the old.

Now, why does the new sum converge? The crucial idea is that we can rewrite each term (except the first, which is unimportant) as a difference of two fractions. We have

$$\frac{1}{2} = 1 - \frac{1}{2},$$
$$\frac{1}{6} = \frac{1}{2} - \frac{1}{3},$$
$$\frac{1}{12} = \frac{1}{3} - \frac{1}{4},$$
$$\frac{1}{20} = \frac{1}{4} - \frac{1}{5},$$

and more generally using algebra we see that

$$\frac{1}{n(n-1)} = \frac{1}{n-1} - \frac{1}{n}.$$

Something very beautiful now happens. When we take the sum of these, most of the terms cancel out! The sum becomes

$$\begin{aligned} 1 + \sum_{n=2}^{\infty} \frac{1}{n(n-1)} &= 1 + \frac{1}{2} + \frac{1}{6} + \frac{1}{12} + \frac{1}{20} + \cdots \\ &= 1 + (1 \underbrace{- \frac{1}{2}) + (\frac{1}{2}}_{} \underbrace{- \frac{1}{3}) + (\frac{1}{3}}_{} \underbrace{- \frac{1}{4}) + (\frac{1}{4}}_{} - \cdots . \end{aligned}$$

But each grouped pair of terms cancels to contribute nothing to the sum, and so the sum is simply $1 + 1 = 2$. (We sometimes describe a sum like this as a *telescoping sum*, because of the neat way that the terms cancel out.) The important thing for us is that this is a finite number, so the sum converges, so the sum of the reciprocals of the squares converges (and in fact must converge to a number less than 2).

Having seen these examples, let's return to our question about how the sum of the reciprocals of the primes behaves. Does it diverge, like the harmonic series, or converge, like the sum of the reciprocals of the squares?

Euler was able to show that the sum of the reciprocals of the primes diverges. It's possible to do this by showing that the initial chunks of the series grow faster than a suitable function, so that the sum diverges: one can show that

$$\sum_{p \leq x} \frac{1}{p} \geq \log \log x$$

and deduce it from that. (The function $\log \log x$ is an increasing function of x: as x gets bigger, so does $\log \log x$, although *very* slowly!)

In particular, a rather weak (but correct) conclusion to draw from the divergence of the sum is that there must be infinitely many primes. After all, if there were only finitely many then the sum would be a sum of finitely many numbers, which must necessarily converge (a sum of finitely many numbers is certainly finite). I say this is a weak conclusion, because we've shown much more than this, but the argument is no less valid. But knowing that the sum diverges gives us a bit more information about the distribution of the primes, which turns out to be rather helpful (it ties in with work on counting primes and the Prime Number Theorem, for example).

This argument of Euler (consider the sum of reciprocals) turns out to extend pretty nicely to showing that there are infinitely many primes with certain properties. In particular, if we want to show that there are infinitely many primes in row a on a d-sided pencil (where a and d have no common factor greater than 1), then we can consider the reciprocals of all such primes, add them all up, and consider the resulting series. It turns out that each such series diverges, and so there must be infinitely many primes on that side of the pencil. The difficult bit is showing that the relevant sum diverges, and so I'm not going to say anything about that here.

Are there more primes that are one more than a multiple of 6 or more that are one less than a multiple of 6?

All this has helped us to resolve some of the questions that the pencil provoked. One question that it hasn't answered is the one about whether there are more primes that are one more than a multiple of 6 or more that are one less than a multiple of 6.

Off the top of my head, I can't think of a reason for a prime to prefer one of those options over the other. So my instinct is that perhaps there's the same number of both (approximately—it seems a bit unrealistic to imagine that the numbers of each type would be *exactly* the same). Looking at computer-generated evidence, it seems that they're fairly evenly divided up to values for which we can check directly. Will that always be the case?

We know how many primes there are up to x: that's the quantity we called $\pi(x)$ before. So we might guess that up to x there are about $\pi(x)/2$ primes that are one more than a multiple of 6, and about $\pi(x)/2$ primes that are one less than a multiple of 6. (Of course, the primes 2 and 3 don't fall into either category, but I'm only really interested in very large x, when $\pi(x)$ is pretty big and so ignoring a couple of primes doesn't make much difference.)

The Elliott–Halberstam Conjecture is a more precise (and technical) version of this prediction—one that averages over pencils with a range of numbers of sides, not just our six-sided pencil. (Since it's rather technical, I'm not going to state the conjecture precisely here.) Mathematicians have proved some results in this direction: for example the Bombieri–Vinogradov Theorem (named after Enrico Bombieri and Ivan Vinogradov) from the 1960s gives us some useful information. It seems plausible that more is true than we've been able to prove so far, and this is the content of the conjecture formulated by Peter Elliott and Heini Halberstam at the end of the 1960s.

The Elliott–Halberstam Conjecture comes with a parameter, conventionally called θ, so really it's a whole family of conjectures, one for each value of θ between 0 and 1. The Bombieri–Vinogradov Theorem tells us that these conjectures are true for any value of θ less than $\frac{1}{2}$ (and hence for all smaller θ too), but mathematicians cannot currently prove the conjecture for any larger value of θ. In fact the situation about comparing numbers of primes on particular sides of a given pencil turns out to be a bit more subtle than it might at first seem, and mathematicians are still seeking a full understanding of the so-called *prime races*.

I'm mentioning the Elliott–Halberstam Conjecture here partly because I think it's interesting (in particular, even the seemingly simple mathematical pencil leads to unsolved research problems), but also crucially because it is relevant to the story of the recent work on the Twin Primes Conjecture. As I mentioned in Chapter 7, Goldston, Pintz and Yıldırım proved that knowing the Elliott–Halberstam Conjecture for any value of θ bigger than $\frac{1}{2}$ would show that there are bounded gaps between primes—but alas nobody has yet found a way to prove exactly what's needed. There has been partial progress: Enrico Bombieri, Etienne Fouvry, John Friedlander and Henryk Iwaniec (in various papers with various combinations of authors) have developed techniques that, under certain technical conditions, prove Elliott–Halberstam-type results for some θ bigger than $\frac{1}{2}$, but unfortunately those technical conditions mean that the work cannot be combined with the argument of Goldston, Pintz and Yıldırım.

Just as an aside, I hope you're getting a sense both of the collaborative nature of mathematics, and also of the incremental way in which each piece of work builds on previous work. In some subjects, past work becomes almost obsolete as understanding develops and new theories disprove old. Mathematics isn't like that, which is why mathematics degrees involve learning a significant amount of nineteenth-century and then twentieth-century mathematics before students are able to reach the cutting edge of twenty-first-century research. (Some aspects of mathematics are more recent and so don't go right back to the nineteenth century, but in other areas the old work is still fundamental to modern mathematics.)

Here's a neat link between the ideas we've been discussing in this chapter and our old friend the Twin Primes Conjecture. That's the conjecture that predicts that there are infinitely many pairs of twin primes, pairs of prime numbers that differ by 2. What happens if we take the sum of the reciprocals of twin primes? That's a series that starts

$$\left(\frac{1}{3}+\frac{1}{5}\right)+\left(\frac{1}{5}+\frac{1}{7}\right)+\left(\frac{1}{11}+\frac{1}{13}\right)+\cdots.$$

Back in 1919, the Norwegian mathematician Viggo Brun (1885–1978) showed that this sum ... drum roll ... converges (its value is now known as *Brun's constant*). If the sum diverged, then we'd have a proof that there are infinitely many twin primes, but it doesn't—so either there are finitely many twin primes, or there are infinitely many and the sum just happens to converge. We can't tell which, but informally it does tell us that there aren't very many twin primes, compared with the number of primes.

Let's get back to the new work on the Twin Primes Conjecture.

CHAPTER 11

August 2013

In a way, the story about August 2013 is that there is no story. This chapter is not going to finish with the exciting news that someone had improved on Zhang's bounds. The Polymath league table did not change during August, or indeed in September. There was no progress to report.

Sometimes that's just how it is in mathematics. Sometimes there is nothing to report. That doesn't mean that nothing is happening, or even that there's no progress, it simply means that there's no *visible* progress.

By this point Polymath had made huge improvements on Zhang's original bound of 70 000 000, managing to show that there are infinitely many pairs of primes that differ by at most 4680. Those improvements came through an interleaved combination of finding better (narrower) admissible sets of a given size, and finding refinements of the theoretical argument to decrease the required size of admissible sets. Inevitably, at some point that dried up—and that point was August 2013. In a way, it's remarkable that the enhancements had continued to come for so long. Certainly the Polymath work was a real achievement.

So what next?

On 17th August 2013, Terry Tao posted on his blog, under the heading 'Polymath8: Writing the paper':

> As has often been the case with other polymath projects, the pace has settled down sub[s]tantially after the initial frenzy of activity; in particular the values ...have stabilised over the last few weeks. While there may still be a few small improvements in these parameters that can be wrung out of our methods, I think it is safe to say that we have cleared out most of the 'low-hanging fruit'

> (and even some of the 'medium-hanging fruit'), which means that it is time to transition to the next phase of the polymath project, namely the writing phase.

He created a skeleton of a possible research paper about the work, and invited others to share the work of writing the paper—and especially to comment on its readability, since it can be difficult for those immersed in a project to gauge what will be comprehensible by those who have thought about it less. The key objective of the paper was to prove the best result of Polymath, namely that there are infinitely many pairs of primes that differ by 4680. But Polymath had other things to share too. For example, the best bounds relied on some very deep and difficult work of the Belgian mathematician Pierre Deligne from the 1970s, but even without using these theorems Polymath had managed to prove that there are infinitely many pairs of primes that differ by at most 14 994, and the techniques that Polymath used to circumvent Deligne's theorems would be of interest to other mathematicians.

In fact it's not quite true to say that the Polymath table of bounds didn't change during August 2013, because this bound of 14 994 was improved very slightly to 14 950 on 17th August, the same day that Tao was posting about writing up the work.

There are some logistical challenges in producing a paper written by many people, but these are not so significant. Potentially more difficult is the matter of how one attributes credit for the work done. As I mentioned in Chapter 7, Gowers addressed this in his first description of the idea of Polymath, when he stipulated that

> Even if only a very small number of people contribute the lion's share of the ideas, the paper will still be submitted under a collective pseudonym with a link to the entire online discussion.

The tradition in mathematics is for all those involved in a project in a major way (and that's typically one or two or three or four people) to appear as authors, in alphabetical order by surname. Other disciplines have their own conventions about who should be named as a first author, or a last author, or listed somewhere in the middle. In mathematics, the convention is that the authors are always listed alphabetically, regardless of who did the majority of the work and who the senior person is. The Polymathematicians created a page on the wiki for the project listing the main contributors (in alphabetical order by surname, naturally) with their affiliations. It is typically a condition of receiving research funding that academics acknowledge such support (and indeed non-financial support) when publishing papers, and this list on the wiki allows that to happen. There is also a list of those who made minor contributions but whose involvement is still significant enough to merit acknowledgement.

The first Polymath papers were submitted under the name DHJ Polymath, with DHJ standing for Density Hales–Jewett (the problem tackled by the first Polymath project). It was agreed in a subsequent Polymath project, prior to bounded gaps between primes, to stick with DHJ Polymath for other projects, rather than changing the initials, not least because it makes it easier to find Polymath papers online.

It's worth noting that the work of writing up this piece of research is not disjoint from doing the research itself. The blog conversations show that some of the discussion was very specifically about writing the paper (comments along the lines of 'I think this definition should come earlier in this section', 'There's a typo at this point', and so on). But much of it was about the mathematical content, about checking the arguments, enhancing the arguments, clarifying the arguments, and so on. Being forced to go through a section slowly and carefully, in order to explain it to others who have not previously thought about it, is a very good way to see whether you really understand it yourself, and to notice mathematical improvements that you had previously overlooked.

By late September 2013, the Polymathematicians (led by Terry Tao) were settling on something like a first draft of their paper, over 160 pages of it. There was still polishing and proofreading to do, but they had reached a consensus on the structure of the paper and the key arguments, and had agreed to submit the paper to the journal *Algebra and Number Theory*. In addition, the editors of the Newsletter of the European Mathematical Society contacted Tao to invite him to write something about Polymath for their publication. He in turn suggested that this should be a collaborative Polymath project too, with participants reflecting on their experience of being involved with Polymath. I'll say more about this in Chapter 16.

On 10th October 2013, Tao posted that he had heard about new work by James Maynard that might improve on the bounds that Goldston, Pintz and Yıldırım obtained by assuming the Elliott–Halberstam Conjecture, based on Maynard's new modification of their sieve. Tao went on to write

> I am beginning to think that there may well be a Polymath8b project in which a number of further improvements beyond the current headline of 4680 are pursued....

Even though there hadn't been much in the way of visible progress in August and September 2013, there were still reasons to be optimistic about imminent new developments....

CHAPTER 12

If primes are hard, let's try something else

I hope that by now you share my feeling that the prime numbers are fascinating, intriguing—and difficult. At least some of the difficulty of the questions we've been considering arises because we are asking *additive* questions about numbers that are defined *multiplicatively*. Prime numbers are described in terms of their (small) number of factors, which is all about multiplication and division, but we've been considering questions to do with addition and subtraction (primes where adding 2 gives another prime, for example). That tension between addition and multiplication makes life interestingly challenging! In this chapter I want to take a little diversion to explore another of my favourite sequences of numbers, also defined multiplicatively, where we can ask similar questions to those we've been asking about the primes, but where it's a bit easier to make progress. This chapter is about the square numbers: $0^2 = 0$, $1^2 = 1$, $2^2 = 4$, $3^2 = 9$, $4^2 = 16$, $5^2 = 25$, and so on. If the primes are the slippery lily pads of the number pond, beautiful and abundant but just a little hard to get hold of, then the squares are the reassuringly solid stepping stones, firmly fixed in place.

Which numbers are sums of two squares?

Goldbach's Conjecture is about which numbers can be written as the sum of two (odd) primes. We could instead ask which numbers can be written as the sum of two squares. One reason to consider this question might be the analogy with

Goldbach's Conjecture. Another might be the link with Pythagoras's Theorem. That's the theorem about right-angled triangles that says that the square of the hypotenuse is equal to the sum of the squares of the other two sides. For the right-angled triangle in Figure 12.1, Pythagoras tells us that $a^2 + b^2 = c^2$.

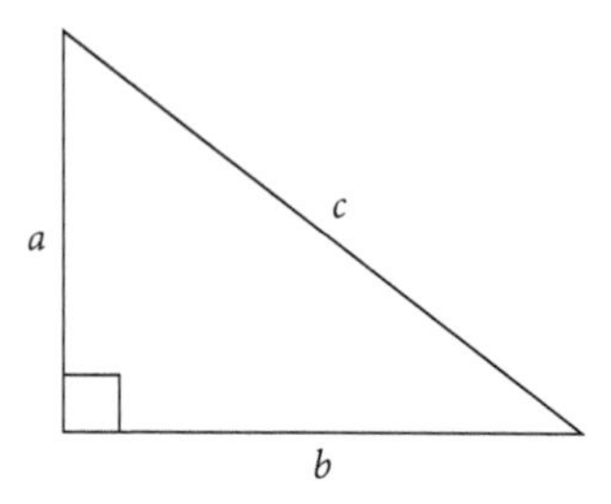

Fig. 12.1 Pythagoras's Theorem: in this right-angled triangle we have $a^2 + b^2 = c^2$.

So thinking about sums of squares of integers corresponds to thinking about possible hypotenuses of right-angled triangles with the short sides having integer lengths. This, for me, is not the main motivation for thinking about sums of two squares, I think the question is interesting in itself (and in particular I'm intrigued by the question, I don't mind whether it has any immediate practical applications!), but there is a bit of a link to triangles. In fact, a fascinating problem in number theory, known as far back as the ancient Greeks, asks for *all* the right-angled triangles with integer side lengths. Authors of school mathematics textbooks are fond of the right-angled triangles with side lengths 3, 4, 5 (note that $3^2 + 4^2 = 5^2$) and 5, 12, 13 (note that $5^2 + 12^2 = 13^2$), because they're nice neat whole numbers and the numbers aren't too big. These

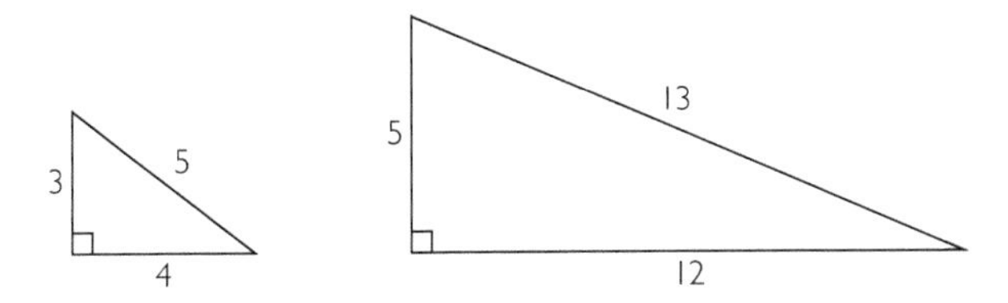

Fig. 12.2 Two right-angled triangles with integer sides.

triangles are shown in Figure 12.2. We say that $(3, 4, 5)$ and $(5, 12, 13)$ are *Pythagorean triples*. But are there more of these Pythagorean triples corresponding to right-angled triangles with integer side lengths? How many? Can we find them all?

It turns out that these questions have beautiful answers. A silly answer to the question 'How many?' is that there are clearly infinitely many, because we can take our favourite starting triangle and scale it up, by multiplying each length by the same amount. If we start with the right-angled triangle with side lengths $(3, 4, 5)$, then we immediately, and for free, get the infinitely many right-angled triangles $(6, 8, 10)$, $(9, 12, 15)$, $(12, 16, 20)$ and so on. This is not an interesting answer! More satisfyingly, it turns out that there are infinitely many genuinely different Pythagorean triples (not just scalings of each other), and in fact the news is as good as it possibly can be. There is an elegant formula that generates all such Pythagorean triples (which I'm not going to give here, but which you might like to think about if you are keen for an interesting investigation of your own).

The problem of finding Pythagorean triples was certainly known to Diophantus of Alexandria (around 200 to around 284), who considered it in his famous work *Arithmetica*. Part of the fame of this book is thanks to one of its later readers: one Pierre de Fermat (1601–1665), a seventeenth century French amateur mathematician who read Claude Bachet's translation—and, happily for us, scribbled his thoughts in the margin of his copy of the book. After Fermat's death, his son published an edition of the *Arithmetica* including his father's annotations. Fermat rarely troubled to explain or justify his reasoning, so other mathematicians went through checking his various assertions. Some turned out to be right, but not all. The last loose end to be tied up became known as Fermat's Last Theorem, and for centuries was one of the most famous unsolved problems in mathematics (giving it the name 'Theorem' was premature!). Fermat did what mathematicians so often do—he sought to generalise.

Diophantus asked about writing a square as a sum of two squares, so, naturally, Fermat wondered about writing a cube as a sum of two cubes, or a fourth power as a sum of two fourth powers, and so on. For example, what are the solutions to the equation $x^3 + y^3 = z^3$, or $x^4 + y^4 = z^4$, or $x^5 + y^5 = z^5$?

Presumably he went away and tried some numerical examples, but all that is lost in the mists of time. All we know is that he somehow convinced himself that while it is possible to write a square number as the sum of two squares, it is never possible to do something similar for powers higher than squares. That is, the equation $x^n + y^n = z^n$ has no solutions for non-zero integers x, y and z if $n \geq 3$, and this was the assertion that he jotted in his copy of the book by Diophantus, along with the provocative remark (in Latin) that

> I have discovered a truly remarkable proof which this margin is too small to contain.

Thus began centuries of mathematical endeavour, as many people tried to prove Fermat's infamous assertion. There were many partial successes along the way, including progress made by mathematicians we have already met in this book such as Leonhard Euler and Sophie Germain.

The problem was finally solved in the mid 1990s, by the British mathematician Andrew Wiles (building on work by numerous others, and also including a crucial component published in a separate paper that was joint work with Richard Taylor, a British mathematician who had been Wiles's graduate student at Princeton). Fermat was right, but it took over 350 years to be sure of this, and it is completely impossible that Fermat had in mind the argument that Wiles later developed—Wiles used modern mathematical machinery that is way beyond anything that was available to Fermat. The story of Fermat's Last Theorem, which is without doubt one of the great stories of mathematics, is full of mistakes, and it seems highly likely that the first of those was Fermat's erroneous belief in his 'remarkable proof'! Don't write off Fermat, though, he'll crop up again in this chapter.

Sums of two squares (continued)

Anyway, back to wondering which numbers can be written as sums of two squares. I want to share with you a good chunk of real mathematics here, going into some of the nitty gritty so that you can get a feel for the kinds of thinking involved. And I've chosen to present it as a story with a punchline at the end, rather than telling you the theorem up front and then presenting a proof, because the process of doing mathematics is about playing with ideas and piecing together information. If you don't feel like thinking about the details now, then you should feel free to skim through the next few pages. You'll find the statement of the theorem and a description of what it tells us at the end of the chapter, and you can always come back to the rest later. But I'm hoping that you'll enjoy getting a flavour of the thinking first!

Let's gather some data, and then we can try to find some patterns and structure.

The square numbers up to 100 are 0, 1, 4, 9, 16, 25, 36, 49, 64, 81 and 100. Figure 12.3 shows them in a grid (except 0).

1	2	3	**4**	5	6	7	8	**9**	10
11	12	13	14	15	**16**	17	18	19	20
21	22	23	24	**25**	26	27	28	29	30
31	32	33	34	35	**36**	37	38	39	40
41	42	43	44	45	46	47	48	**49**	50
51	52	53	54	55	56	57	58	59	60
61	62	63	**64**	65	66	67	68	69	70
71	72	73	74	75	76	77	78	79	80
81	82	83	84	85	86	87	88	89	90
91	92	93	94	95	96	97	98	99	**100**

Fig. 12.3 Square numbers.

Any square number is certainly a sum of two square numbers (for example, $25 = 5^2 + 0^2$). Which extra numbers do we get by adding non-zero squares? Figure 12.4 shows the numbers that are sums of two squares, with the squares themselves in a different colour (but these still count as sums of two squares).

Do you remember the beautifully clear pattern that we found when carrying out this exercise for sums of two primes? Alas, no such clear pattern here with sums of two squares. But.... Why should we have ten columns? Mostly because humans have ten fingers (well, digits). But what if we use say eight columns instead? That's shown in Figure 12.5.

That's a bit more like it! There are definitely some interesting observations to be made here.

One is that three of the columns appear to contain no sums of two squares at all (the columns starting 3, 6 and 7). Is that really the case? Will that continue if

1	2	3	4	5	6	7	8	9	10
11	12	13	14	15	16	17	18	19	20
21	22	23	24	25	26	27	28	29	30
31	32	33	34	35	36	37	38	39	40
41	42	43	44	45	46	47	48	49	50
51	52	53	54	55	56	57	58	59	60
61	62	63	64	65	66	67	68	69	70
71	72	73	74	75	76	77	78	79	80
81	82	83	84	85	86	87	88	89	90
91	92	93	94	95	96	97	98	99	100

Fig. 12.4 Sums of two squares (with squares in blue).

1	2	3	4	5	6	7	8
9	10	11	12	13	14	15	16
17	18	19	20	21	22	23	24
25	26	27	28	29	30	31	32
33	34	35	36	37	38	39	40
41	42	43	44	45	46	47	48
49	50	51	52	53	54	55	56
57	58	59	60	61	62	63	64
65	66	67	68	69	70	71	72
73	74	75	76	77	78	79	80
81	82	83	84	85	86	87	88
89	90	91	92	93	94	95	96
97	98	99	100				

Fig. 12.5 Sums of two squares (with squares in blue).

we go further down the columns? And how can we predict which of the numbers in the other five columns are sums of two squares and which aren't, is there a neat way to work this out (without lots of calculations)? As mathematicians, we are always looking for patterns and structure, and then trying to explain the reasons behind the patterns and structure.

It will be helpful to start by looking at where the squares are in our eight-column grid (this is why I left them in a different colour even though they are themselves sums of two squares). From the evidence of the squares up to 100, it looks as though each square is in one of three columns (the 1, 4 and 8 columns).

What does it mean to be in the 8 column? These numbers are precisely the multiples of 8. The numbers in the 1 column are the numbers that are 1 more than a multiple of 8, and similarly the numbers in the 4 column are the numbers that are 4 more than a multiple of 8.

So now the key question is whether it's really true that every square is a multiple of 8 or 1 more than a multiple of 8 or 4 more than a multiple of 8. There are various ways to see that this is the case.

It turns out that one useful approach is to split the whole numbers into four categories: multiples of 4, numbers that are 1 more than a multiple of 4, numbers that are 2 more than a multiple of 4, and numbers that are 3 more than a multiple of 4.

Algebraically, a multiple of 4 can be written as $4k$ where k is an integer. A number in the second category can be written as $4k+1$ where k is an integer, and similarly for the remaining two cases. Then the squares of these numbers are

$$\begin{aligned}(4k)^2 &= 16k^2 = 8 \times 2k^2;\\ (4k+1)^2 &= 16k^2 + 8k + 1 = 8(2k^2 + k) + 1;\\ (4k+2)^2 &= 16k^2 + 16k + 4 = 8(2k^2 + 2k) + 4; \text{ and}\\ (4k+3)^2 &= 16k^2 + 24k + 9 = 8(2k^2 + 3k + 1) + 1.\end{aligned}$$

This shows that every square in the world is a multiple of 8, or 1 more than a multiple of 8, or 4 more than a multiple of 8. This turns out to be a rather useful fact for all sorts of problems. It also contains the amusing snippet that the square of every odd number is 1 more than a multiple of 8, which is perhaps rather surprising!

(It might seem more natural to think about eight cases, depending on the remainder when the number is divided by 8, and this also works out just fine. I have applied the well known mathematical tool of hindsight to give the streamlined four cases above!)

What does this tell us about sums of two squares? Well, it means that every sum of two squares is a sum of two numbers that are 0, 1 or 4 more than a multiple of 8. To think about the possible sums, we can simply concentrate on these remainders. For example, the sum of a multiple of 8 and a number that is one more than a multiple of 8 is $0 + 1 = 1$ more than a multiple of 8. So we can just go through the possibilities, since there aren't very many of them:

$$0 + 0 = 0$$

$$0 + 1 = 1$$

$$0 + 4 = 4$$

$$1 + 1 = 2$$

$$1 + 4 = 5$$

$$4 + 4 = 0.$$

Note how I've been systematic about this, I've checked them in a logical order, to make sure that I checked all the cases, but also I was able to reduce the number by realising that the order in which we add the numbers isn't important: I didn't need to check $1 + 0$ because it's the same as $0 + 1$, which *is* on the list.

The important thing here is not so much which remainders appear on the right-hand side as which *don't*. Our list above reveals that no sum of two squares ever leaves remainder 3, 6 or 7 when divided by 8, and that confirms that the 3, 6 and 7 columns in Figure 12.5 do not contain any sums of two squares, no matter how far down the columns we go. Isn't that nice?! I love the certainty, I love that such an elegant argument can unambiguously guarantee that there are no sums of two squares in those columns.

What it can't do, however, is tell us which numbers in the remaining columns *are* sums of two squares. It's clear from Figure 12.5 that some are and some aren't, and the argument with remainders tells us no more. It's certainly not clear to me what the patterns are here, based on looking at the grid. So we need another idea. Solving mathematics problems is like this: an idea helps you to make a bit of progress, and then you start to detect where the complexities lie and where you need some new ideas or approaches.

Multiplying sums of two squares

Here's an observation that will turn out to be very useful. If we take two numbers that are both sums of two squares, then multiplying them together gives a number that is also a sum of two squares. We can test the claim on numerical examples from earlier, for example

$$5 = 2^2 + 1^2 \quad \text{and} \quad 13 = 3^2 + 2^2,$$

so 5 and 13 are both sums of two squares, and indeed

$$5 \times 13 = 65 = 4^2 + 7^2$$

so their product is also a sum of two squares. That's not a proof, but we could use it to try to look for some patterns. It's a little bit fiddly to use this sort of numerical data to guess the relationship between the squares used for the first two numbers and the squares used for the product, made even more difficult because there could be more than one way to write the numbers as sums of two squares (for example, $65 = 4^2 + 7^2$ but also $65 = 8^2 + 1^2$). Either one has to work quite hard to look at the numerical evidence to try to find patterns, or one needs another way to sneak up on the problem (I'll say more about this in a moment).

It turns out that the key is the algebraic identity

$$(a^2 + b^2)(c^2 + d^2) = (ac - bd)^2 + (ad + bc)^2,$$

which holds for all values of a, b, c and d.

Let's think about the example above. We had

$$5 = 2^2 + 1^2,$$

so we could think about $a = 2$, $b = 1$.

We also had

$$13 = 3^2 + 2^2,$$

so let's take $c = 3$ and $d = 2$.

Then $ac - bd = 6 - 2 = 4$ and $ad + bc = 4 + 3 = 7$ and we recover the first expression of

$$65 = 4^2 + 7^2$$

as a sum of two squares from above.

Why does the general equation hold? Well, one way to check is to multiply out the brackets on each side of the equation, and check that the resulting two expressions are equal. That's entirely satisfactory to confirm the truth of the identity, but entirely frustrating in that it gives no insight into where the equation comes from. How might we have thought of it in the first place (other than guessing based on numerical patterns)?

One rather nice answer is 'via complex numbers'. Complex numbers aren't necessary to check that the identity is true, but they do explain very neatly why the identity is a sensible thing to write down in the first place. I'm not going to go into lots of detail about what complex numbers are here, so if you don't know about them then just skip over the next paragraph! I'll say a bit more later in the chapter, though.

The point is that $a^2 + b^2$ is the square of the length of the complex number $a + bi$, and $c^2 + d^2$ is the square of the complex number $c + di$. When we multiply those two complex numbers, we get

$$(a + bi)(c + di) = (ac - bd) + (ad + bc)i,$$

whose length squared is $(ac - bd)^2 + (ad + bc)^2$. So the mysterious right-hand side of the identity suddenly arises from a very natural source!

Why is this identity so helpful? We're trying to establish which numbers can be written as a sum of two squares, so why is it so great to know that if two numbers are sums of two squares then so is their product? Well, it means that looking at *prime* numbers is likely to be a good idea, because they can be used as building blocks to make up all other numbers. I mentioned this briefly in Chapter 2, where the idea was encapsulated in the dotty diagrams of numbers showing their prime factorisations. Let me say more about this now, and then we'll return to sums of two squares in a moment.

Interlude on prime factorisations

Each whole number bigger than 1 has a prime factor: either the number itself is prime, or it's divisible by a smaller prime. Pursuing that thought, repeatedly pulling out prime factors, we find that each number can be written as a product of prime numbers (a bunch of prime numbers multiplied together). For example, $18 = 2 \times 3 \times 3$ and this is a product of prime numbers. (It doesn't matter that the 3 is repeated, that's allowed.) And 17 is prime so is a product of prime numbers—albeit a rather dull product consisting of just the single number 17.

Amazingly, it turns out that there's only one way to do this for each number. If you like, there's only one possible diagram for each number. The diagram depends on the order in which we think of the factors: the diagram for 2 lots of 3 lots of 3 would look different from the diagram for 3 lots of 3 lots of 2 (see Figure 12.6).

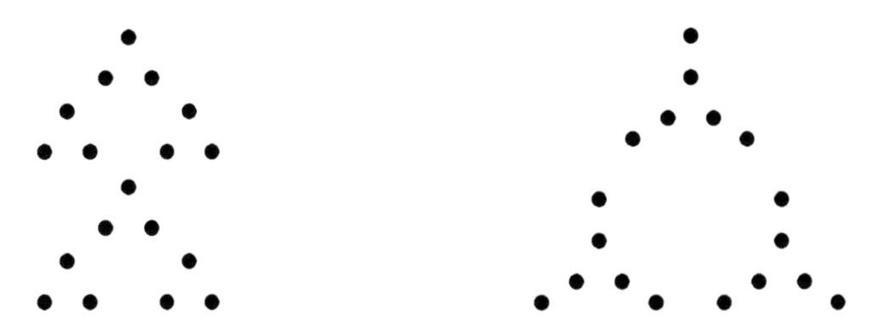

Fig. 12.6 2 × 3 × 3 and 3 × 3 × 2 lead to different diagrams.

However, if we agree that we always start with the biggest factor and work down when arranging the dots, then there's only one diagram. This property is called the *uniqueness of prime factorisation*, and it's one of the things that makes prime numbers so special. Not only is it possible to break every number down into its prime factors, but also there's only one way to do this for each number (apart from writing the factors in different orders). This uniqueness is not obvious: our experience of examples may suggest that it's true, but that's not good enough, it needs a proof. I'm not going to give a proof right now, so please take my word for it—or, better still, you could look up a proof online or in a number theory textbook from the list of Further Reading at the end (or, even better, think of a proof yourself!). The theorem is called the Fundamental Theorem of Arithmetic; the name alone conveys the importance of this result.

Theorem (Fundamental Theorem of Arithmetic) *Every whole number greater than 1 can be expressed as a product of primes in an essentially unique way.*

I've written 'essentially unique' because there's the slight wrinkle that the factors can be written in a different order, but this doesn't count as a different factorisation. I'm a pure mathematician, so I try very hard to make precise statements!

By the way, this is a good reason not to allow 1 to be a prime number. If 1 were prime, then we wouldn't have uniqueness of prime factorisation: the factorisations $2 \times 3 \times 3$ and $1 \times 1 \times 2 \times 3 \times 3$ would technically be different, and that would not be helpful.

The Fundamental Theorem of Arithmetic tells us two important things about prime factorisations. One is that every whole number greater than 1 *has* a prime factorisation. That's important, but not subtle. We can see why it's true quite quickly. Pick a number. If it's prime, then we're done. If not, then it's a product of smaller numbers. By repeating the same argument on each of those, we eventually find a prime factorisation of our original number.

For example, we might pick 120, and then factorise it and then factorise those factors and so on:

$$\begin{aligned} 120 &= 10 \times 12 \\ &= 2 \times 5 \times 2 \times 6 \\ &= 2 \times 5 \times 2 \times 2 \times 3. \end{aligned}$$

Not only do we know that a prime factorisation exists, we even have a strategy for finding one (albeit one that is potentially *extremely* time-consuming, if our starting number is enormous). The second, more subtle, part of the theorem is the uniqueness. Proving that is a bit more delicate, because it relies on deeper properties of prime numbers, and this is the part that means that the theorem is not obvious.

The significance of this became clear in the nineteenth century, when mathematicians started expanding the types of numbers they were working with. Gabriel Lamé (1795–1870) had a brilliant plan of attack for Fermat's Last Theorem, building on a conversation he'd had with Joseph Liouville (1809–1882)—in fact Lamé announced that he had a proof, which involved using not just the usual integers, but an extended pool of numbers obtained by including some complex numbers. His argument involved factorising $x^n + y^n$ and drawing conclusions about what would happen if this equalled the nth power z^n to show that there are no solutions. Ingenious though this was, unfortunately it doesn't work for some values of n, because, as Liouville pointed out, it relies on uniqueness of prime factorisation and this *fails* for some cases in these extended pools of numbers. At this point there might have been a temptation to give up, but in fact at least a hundred years of modern algebraic number theory developed out of this disappointment, starting with the work of Ernst Eduard Kummer (1810–1893), trying to get round the failure of the uniqueness of prime factorisation (which in fact led to a correct proof of Fermat's Last Theorem for some exponents). Mathematicians make mistakes all the time, the secret is to use them to build our understanding!

One of the messages here is that uniqueness of prime factorisation in the integers is a wonderful thing, and we should not take it for granted.

Which (prime) numbers are sums of two squares?

Now back to sums of two squares. We have seen that each whole number bigger than 1 can be obtained by multiplying together suitable prime numbers, and we have seen that multiplying sums of two squares leads to a number that is a sum of two squares. So if we could understand which *prime* numbers are sums of two squares, then perhaps we could piece together that information to understand the more general situation about sums of two squares. This strategy, of understanding the behaviour of prime numbers and then piecing together the resulting information to deduce the behaviour of all whole numbers, is a classic one: it's one of the many reasons why the Fundamental Theorem of Arithmetic is so important.

Let's have another look at that grid showing sums of two squares, this time also highlighting the primes. In Figure 12.7, the sums of two squares are coloured yellow, the primes are coloured blue, and so naturally the green numbers are those that are both prime and a sum of two squares.

1	2	3	4	5	6	7	8
9	10	11	12	13	14	15	16
17	18	19	20	21	22	23	24
25	26	27	28	29	30	31	32
33	34	35	36	37	38	39	40
41	42	43	44	45	46	47	48
49	50	51	52	53	54	55	56
57	58	59	60	61	62	63	64
65	66	67	68	69	70	71	72
73	74	75	76	77	78	79	80
81	82	83	84	85	86	87	88
89	90	91	92	93	94	95	96
97	98	99	100				

Fig. 12.7 Yellow sums of two squares, blue primes, and green primes that are sums of two squares.

If we focus just on the green and blue numbers, there are some pretty striking patterns there. As so often happens, 2 is a special case, so let's put it to one side—we know that 2 is a sum of two squares, so we can quietly ignore it for the time being. What about the odd primes? Well, looking at that grid it seems pretty clear: primes in the 3 and 7 columns are not sums of two squares, and primes in the 1 and 5 columns are (at least for the numbers in the grid). If that's true for all primes, not just those in the grid, it would be brilliant, because it's easy to check which column a prime is in—just divide by 8 and check what the remainder is—and then that would immediately tell us whether the prime is a sum of two squares.

We know the answer to half of this already. We carefully checked above that *no* number in the 3 or 7 column is a sum of two squares, and this certainly applies to the prime numbers in those columns. The primes in the 3 and 7 columns are all blue, there can't be any green.

What about the other half? Is it the case that *every* prime that is in the 1 or 5 column can be written as the sum of two squares? That feels harder to prove.

You might have noticed that we could streamline our descriptions slightly. Rather than thinking about remainders on division by 8, we could think about remainders on division by 4. The 1 and 5 columns in our grid contain precisely the numbers that leave remainder 1 when divided by 4, and the 3 and 7 columns contain the numbers that leave remainder 3 when divided by 4. (If you like, you could redraw the grid with four columns to illustrate this. Apart from the 2, the green numbers—the primes that are sums of two squares—would all be in column 1, and the blue numbers—the primes that are not sums of two squares—would all be in column 3.) So we know that any number that's 3 more than a multiple of 4 is not a sum of two squares, that's what we checked above. Now we'd like to know whether every prime number that's 1 more than a multiple of 4 can be written as a sum of two squares.

This turns out to be true, and in fact it's a theorem attributed to Fermat.

Theorem (Fermat) *The odd prime number p can be expressed as a sum of two squares if, and only if, p is 1 more than a multiple of 4.*

True to form, Fermat claimed the result in a letter but neglected to share a proof with anyone. It took just over a hundred years for the first known proof, which Euler gave in a letter to Goldbach (see how the same characters keep cropping up in these stories?). Nonetheless, it tends to be known as Fermat's theorem. Since Euler's proof, there have been a number of other gorgeous proofs using various different ideas from the theory of numbers.

I'll sketch an outline of an argument here, but you can skip over it if you want. I've chosen this approach because it's linked to the (sort of) failed attempt on Fermat's Last Theorem in the interlude on prime factorisations above: rather than working just with the integers, we expand our pool of available numbers to include some complex numbers, and use them for our argument.

This is really not the moment for an introduction to the theory of complex numbers, but let me see whether I can tell you just enough for you to get a flavour of the proof. To start with, we need to allow ourselves to use i, the square root of -1. This seems like a pretty terrifying thing, because the square of any usual number (*real number*), such as 2 or -17 or π or -0.01, is greater than or equal to 0. So how can -1 have a square root? Well, somehow we shouldn't get hung up on this. All we need to do is to invent a new symbol, i, and remember that every time we see i^2 we can replace it by -1. I am here conveniently glossing over several hundred years of mathematical thought, but after all it took several hundred years to decide that it was safe to use 0, and also to use negative numbers, and we now use those all the time!

We're going to work with the *Gaussian integers*, named after the great German mathematician Carl Friedrich Gauss (1777–1855). These are numbers of the form $a + bi$ where a and b are just usual integers (whole numbers). For example, $1 + 2i$, $-1 + 3i$, $7 - 6i$ and $-8 - 2i$ are all Gaussian integers. We can add, subtract and multiply Gaussian integers in much the same way that we add, subtract and multiply the usual integers, remembering always that if we see i^2 then we can replace it by -1. For example,

$$(1 + 2i) + (7 - 6i) = (1 + 7) + (2 - 6)i = 8 - 4i,$$

$$(1 + 2i) - (7 - 6i) = (1 - 7) + (2 - -6)i = -6 + 8i, \text{ and}$$

$$(1 + 2i)(7 - 6i) = 7 + 14i - 6i - 12i^2$$

$$= (7 + 12) + (14 - 6)i = 19 + 8i.$$

I want to mention one more thing about Gaussian integers. To each Gaussian integer $a + bi$, we assign a quantity $L(a + bi)$, which we define to be $a^2 + b^2$. Notice that this is always a normal integer, because a and b are normal integers. (If you know about complex numbers, then you'll know that it's possible to interpret them geometrically, and then to assign a length to each complex number, and $L(a + bi)$ is the square of the length of $a + bi$.)

If you're keen, you could look back a few pages to that identity that shows that the product of two sums of two squares is also a sum of two squares, to revisit that argument in light of this experience with Gaussian integers. Using our new notation, the identity records the fact that $L(a + bi) \times L(c + di) = L((a + bi)(c + di))$ (we can do L to the two complex numbers and then multiply the answers, or we can multiply the numbers and then do L, we get the same answer either way round). This will be useful to us again in a moment.

Now, here's a neat observation. If we work over the usual integers, 29 is prime and so has no interesting factorisations. Things like $(-1)\times(-29)$ count as boring factorisations! (Sorry if 'over the usual integers' seems like strange phrasing, but it's just what mathematicians say!) However, over the Gaussian integers, 29 does factorise:

$$29 = (5 + 2i)(5 - 2i) = 5^2 + 2^2.$$

This factorisation over the Gaussian integers corresponds to an expression of 29 as a sum of two squares! This is part of the idea behind this proof of Fermat's theorem.

The other part, which I'm not going to prove here, is that if p is a prime that is 1 more than a multiple of 4, then p divides a number $m^2 + 1$ for some suitable integer m. For example, if $p = 5$ then we can choose $m = 7$ because 5 divides $7^2 + 1 = 50$, and if $p = 29$ then we can choose $m = 12$ because 29 divides $12^2 + 1 = 145$. That this holds in general is very definitely not obvious, but it requires a bit more theory than I can include right now, so I hope you'll be happy just to believe it (or you could try to prove it for yourself).

Let's think about this for a bit. This argument is getting quite challenging now, but we are nearly there. I'm going to press on, because even if you don't follow all the details right now, you can safely skim over it and return to it later if you're interested. This is what mathematicians do absolutely all the time when reading papers and books. It is often the case that reading mathematics in a linear fashion is not the best approach. Rather, we read the introduction to get an idea of what it's all about, we flick to the bits that look juicy, we turn back to an earlier part because it turns out that the impenetrable section 3.7.2 (or whatever) matters after all, and then we spend time deciphering section 3.7.2 because we now know why we care, namely that it leads into the really interesting and innovative part of the proof of Theorem 2. Maybe we can come up with a nicer argument for section 3.7.2 ourselves if we get stuck in. Anyway, back to sums of two squares, there's a page or so more of this stuff, and then we'll get to the punchline, which I've labelled with a section heading in case you're wanting to race ahead.

We know that 5 divides $7^2 + 1$. Over the Gaussian integers, we can factorise this: $7^2 + 1 = (7 + i)(7 - i)$. Like the usual integers, the Gaussian integers have the property of unique prime factorisation, so we're allowed to talk about 'the prime factorisation' of a number.

The prime factorisation of $7^2 + 1$ is simply the product of the prime factorisations of $7 + i$ and $7 - i$. Neither of these latter two contains 5 (if we divide $7 + i$ by 5 then we get $\frac{7}{5} + \frac{i}{5}$, which is not a Gaussian integer), so 5 does not appear in the prime factorisation of $7^2 + 1$. But 5 certainly divides it, so 5 cannot be prime in the Gaussian integers: it must have an interesting factorisation as $5 = (a + bi)(c + di)$ for some (normal) integers a, b, c and d. (We could of course write down this factorisation, we have $5 = (2 + i)(2 - i)$, but I'm going to stick with the algebra because I want to have a general sort of argument that we can extend to numbers other than 5.)

Now let's think about applying L to this factorisation of 5. The great thing about L is that it takes mildly scary Gaussian integers and turns them into friendly and familiar normal integers. Using our identity from earlier, we see that $L(5) = L(a + bi)L(c + di)$. In particular, $L(a + bi)$, which is just some integer, must divide $L(5)$, which is 5^2. But a consequence of the factorisation being *interesting* is that neither $L(a + bi)$ nor $L(c + di)$ is 1, and so the only possibility for $L(a + bi)$ is that it is 5. That is, $5 = L(a + bi) = a^2 + b^2$—and so 5 is a sum of two squares!

We really didn't need all this machinery to discover that $5 = 2^2 + 1^2$ is a sum of two squares, but if we go through the last two paragraphs and replace every 5 by a general p, and every 7 by the appropriate m (one with the property that p divides $m^2 + 1$) then we'll have a general argument. Bingo!

Where did we use the fact that p is 1 more than a multiple of 4? Remember that we're really expecting to use that, because we know that the statement is *false* for other odd primes p. It was crucial in knowing that there's an m so that p divides $m^2 + 1$—that statement is not true otherwise.

If you look up some of the references in the Further Reading at the end, or just search online, you'll find other approaches to this proof, but I hope that this one has given you a flavour of how introducing complex numbers can be useful for solving problems about whole numbers.

Sums of two squares: the conclusion

We now have a lovely way to determine whether a prime number is a sum of two squares: Fermat's theorem tells us that we can simply find its remainder on division by 4 (or 8). How can we use this to find a neat way to decide whether *any* number is a sum of two squares?

Let's pick any old number, let's call it n, and let's focus on its prime factorisation. We've seen that if two numbers are both sums of two squares, then so is their product, so the prime factorisation of n is going to be helpful for learning whether or not n is a sum of two squares. I find myself thinking of the prime factors of n as being in three piles (which are allowed to be empty): the primes that are 1 more than a multiple of 4, the primes that are 3 more than a multiple of 4, and the prime 2 (if n is even). Figure 12.8 shows a couple of examples.

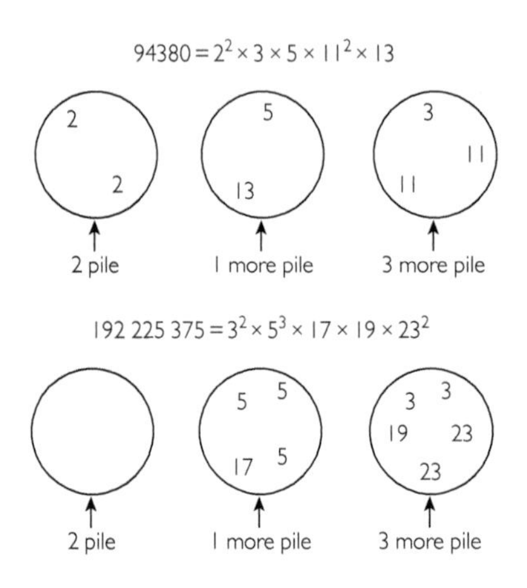

Fig. 12.8 Piles of prime factors.

The prime 2 is good news for us, because it is a sum of two squares. Any prime factor that's 1 more than a multiple of 4 is good, because it's a sum of two squares (by Fermat's theorem above). The *square* of any prime factor that's 3 more than a multiple of 4 is good, because any square number is a sum of two squares and so good. We can safely ignore all these good factors, because their product is a sum of two squares. So we can discount the 2 pile, and the pile of primes that are 1 more than a multiple of 4, and in the pile of primes that are 3 more than a multiple of 4, we can remove pairs that are the same. I've illustrated this in Figure 12.9.

For example, $23465 = 5 \times 13 \times 19^2$. Here, 5 is good, and 13 is good, and 19^2 is good, and so 23 465 must be a sum of two squares. To write it explicitly in this form, we can use the formula from earlier to handle a product of sums of two squares (or just use a computer to test all the possibilities, since 23 465 is relatively small). Earlier in this chapter we saw that $5 \times 13 = 4^2 + 7^2$, and so $23\,465 = 5 \times 13 \times 19^2 = (4 \times 19)^2 + (7 \times 19)^2 = 76^2 + 133^2$.

What we still need to worry about is any leftover primes that are 3 more than a multiple of 4. There is only one of each of these (if there were two or more, then we could remove one or more pairs). For example, what about a number

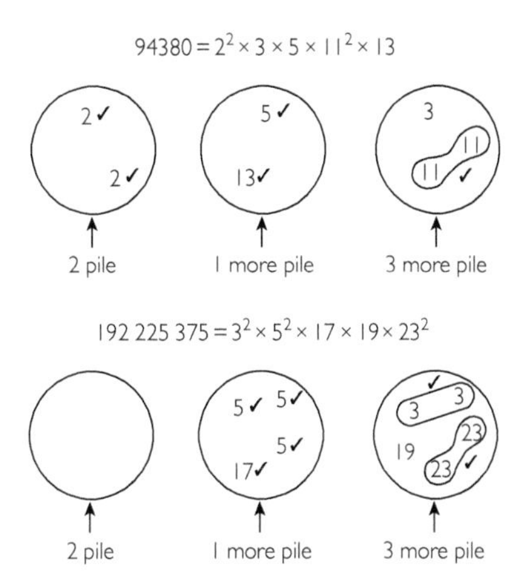

Fig. 12.9 Piles of prime factors showing which are good.

like $7 \times 23 = 161$? Here each of the prime factors is 3 more than a multiple of 4. A check of the possibilities shows that 161 is *not* a sum of two squares, but why is this?

To deal with this, we're going to reuse some of the ideas from earlier. (Mathematicians are good at recycling! Why have new ideas if you can reuse ones you've already had?) Let's take a sum of two squares, say $x^2 + y^2$, and let's take a prime p that divides $x^2 + y^2$ and that is 3 more than a multiple of 4. Our aim is to show that p *must* appear to an even power in the prime factorisation of $x^2 + y^2$—that will explain cases such as 7×23 above, cases where a prime factor that is 3 more than a multiple of 4 appears to an odd power.

We're going to use the Gaussian integers again. (This means that if you didn't follow the previous argument then you might want to skim over the next couple of paragraphs, or you might want to flick back a couple of pages to revisit the previous discussion!).

Is p still prime over the Gaussian integers, or could it factorise? Well, just as we saw for the example of 5 previously, if it factorises interestingly then it's $p = (a + bi)(c + di)$ for some integers a, b, c and d with $L(a + bi)$ and $L(c + di)$ both not 1. Also we have $L(p) = p^2$ and $L(p) = L(a + bi)L(c + di)$, so we must have $p = L(a + bi) = a^2 + b^2$—that is, p is a sum of two squares. But we know that p is not a sum of two squares because it's 3 more than a multiple of 4, we saw that several pages back. So p cannot have an interesting factorisation over the Gaussian integers.

Now back to our sum of two squares. We can factorise $x^2 + y^2$ as $(x + yi)(x - yi)$, and we know that this is divisible by p. But p doesn't have an interesting factorisation, so it must divide one of the factors: it must divide $x + yi$ or $x - yi$. When we think about this for a moment, we realise that it means that p must divide both of x and y. So now we have $x = pX$ and $y = pY$ for some (normal) integers X and Y, and our original sum of squares factorises as $x^2 + y^2 = p^2(X^2 + Y^2)$. So we've pulled out an even power of p. If p still divides $X^2 + Y^2$, then we're going to be able to repeat the argument to pull out another p^2, and, by repeating this enough times, eventually we'll be left with a sum of squares that's not divisible by p. So in the prime factorisation of the original sum of two squares, p must appear to an even power. To turn that around, if p appears to an odd power then the number cannot be a sum of two squares.

We've (more or less) proved a theorem!

Theorem *The positive integer n can be written as a sum of two squares if, and only if, in the prime factorisation of n every prime that is 3 more than a multiple of 4 appears to an even power.*

(The 'more or less' is because there were one or two parts where I didn't fill in the details, not because I said anything untrue on the way!)

Putting that another way, we have an easy method to check whether a number is a sum of two squares. Pick your number, work out its prime factorisation, look for the primes that are 3 more than a multiple of 4, and see whether they all appear to even powers.

For example, $68 = 2^2 \times 17$. There are no primes that are 3 more than a multiple of 4 here, so 68 is a sum of two squares (for example, $68 = 8^2 + 2^2$).

By way of contrast, $98 = 2 \times 7^2$. Here 7 is a prime factor and is 3 more than a multiple of 4, but it occurs to an even power so we're safe—98 is a sum of two squares (it's $7^2 + 7^2$, in fact).

A third example: $99 = 3^2 \times 11$. Here 3 and 11 are both prime factors that are 3 more than a multiple of 4, and while 3 is well behaved (it appears to an even power), 11 is not, and so 99 is *not* a sum of two squares.

For me, the conclusion (the theorem and associated method for checking whether a number is a sum of two squares) is immensely satisfying, but I love the journey just as much. Even if you just skimmed through it very quickly, I hope that you might have got a little bit of the flavour of the argument.

In particular, this highlights that, at least in this respect, squares are much easier to understand than prime numbers are! We can prove a precise theorem that tells us which numbers are sums of two squares, whereas the corresponding assertion for prime numbers (Goldbach's Conjecture), that every even integer bigger than 2 is a sum of two primes, remains an open problem.

On the subject of unsolved problems, let's check back in with the work on bounded gaps between primes to see how that's progressing . . .

CHAPTER 13

November 2013

'Fortune favours the prepared mind' said Louis Pasteur in the 1850s. This is as true in mathematics as it is in other disciplines. When Zhang announced his big breakthrough, there was one person particularly well placed to appreciate the significance of his work. James Maynard had recently completed his PhD work in Oxford, under the supervision of Roger Heath-Brown (who made an appearance in Chapter 7 for his work in the 1980s on the Twin Primes Conjecture). He was spending some time in Montreal, as a postdoc working with Andrew Granville, himself a leading number theorist. As a graduate student, Maynard had studied techniques from analytic number theory, including sieve theory. When Zhang's work appeared, Maynard was already an expert on the related work of Goldston, Pintz and Yıldırım, so it took him much less time to appreciate and assimilate Zhang's work than it took some of the other mathematicians who got involved in the project to improve the bounds.

When Zhang's paper came out, Maynard might have been disappointed: perhaps he would himself have been able to prove bounded gaps between primes before too long—he was certainly working on the right sorts of problems, and had made significant contributions to the area. In fact, he recognised the benefit that Zhang's work had for him, namely making a relatively quiet backwater of mathematics into a seriously hot topic, one of interest not only to other number theorists, but also to a much wider audience. These things do matter: whether rightly or wrongly, a mathematician working in a fashionable research area gets their work noticed more, gets more invitations to speak at conferences, and stands more chance of being given jobs and prizes.

By the time the Polymath project on bounded gaps between primes began, Maynard had already studied Zhang's work in detail, and was making good progress with developing his own ideas. Within the mathematical community, there are numerous ways in which people think about their work. Some mathematicians approach their research by tackling particular problems that intrigue them. Others prefer to focus on finding common features in seemingly disparate areas of mathematics and then abstracting out those common features to develop new theory. Maynard describes his chosen way of working as being about exploring techniques and approaches, seeking to understand their limitations. How far can methods from sieve theory take us? Maynard wanted to test the boundaries of what's possible, by combining his own ideas with those of Goldston, Pintz and Yıldırım and with those of Zhang, to see how far he could get. Since he did not expect to make a major breakthrough, he was not in a race against Polymath, but thought that perhaps he could come up with something that might contribute to the Polymath effort. He preferred to work on his own ideas rather than to join the massive collaboration, but nonetheless checked on Polymath progress from time to time, to find out how they were getting on.

And then in the autumn of 2013 he realised that his ideas had come together much more successfully than he had imagined would be possible: Maynard could prove a much better bound than even Polymath had obtained. At about this time, he learned that Terry Tao had independently had very similar ideas. As we've seen elsewhere in this story (for example Hadamard and de la Vallée Poussin proving the Prime Number Theorem), this does happen, and it can be tricky for the individuals involved. The dispute between Newton and Leibniz over who first thought of the ideas of calculus is legendary. On this occasion, there was some tactful negotiation, helped by Andrew Granville, who was Maynard's mentor and who knew Tao well, and it was agreed that Maynard would publish his work, with a full acknowledgement that Tao had independently proved a similar result, and that Tao would describe his own work on his blog. This was generous on the part of Tao (who could have written up his work for publication too), and might stem from an acknowledgement that the credit for the progress meant more to Maynard, right at the start of his career, than to such an established mathematician as Tao. In addition, Maynard's version of the argument yielded better bounds than Tao's.

Both versions (Maynard's paper on the arXiv and Tao's post on his blog) appeared on 19th November 2013, only a few months after Zhang's original announcement. As often happens, news of the breakthrough had been circulating before that. On 18th October Tao had posted a comment in a Polymath thread (on his own blog) hinting that he had heard that Maynard could get better bounds, but at that stage the announcement was provisional:

> this is all numerical work and is subject to confirmation. The details are not yet written up and probably won't be for a few more weeks yet.

A few days later, on 23rd October, news was reaching Polymath from Oberwolfach that Maynard could prove a bound of 700 (that is, there are infinitely many pairs of primes that differ by at most 700). The Mathematisches Forschungsinstitut Oberwolfach, or Oberwolfach Research Institute for Mathematics, known amongst mathematicians simply as Oberwolfach, is a major international mathematics institute tucked away in the Black Forest in Germany. Experts are invited to gather there for intensive workshops in their research areas to hear about each other's work and to develop and continue international collaborations. Maynard was in Oberwolfach for a workshop on Analytic Number Theory, with a group including both established experts and members of the coming generation of researchers in the area, and his announcement was just one of many that week in Oberwolfach.

As news got out, the Polymathematicians started wondering what exactly Maynard had done: the rumours were that his bounds were very good, and that his argument was easier than previous ones. One participant, the Hungarian mathematician Gergely Harcos, wrote on Tao's blog

> All this is very exciting indeed. It is always a good sign when a proof simplifies and the result becomes stronger simultaneously.

The news of Maynard's work was breaking as the Polymathematicians were writing up their work to submit to a journal, so there was some discussion about how best to proceed now that their work was no longer the world record. They still had some interesting new ideas to share, so it was important to write up their work for official publication, and that process continued during the few weeks it took for Maynard and Tao to present their respective work online for scrutiny by others.

What had Maynard and Tao proved? Excitingly, they were able to obtain two types of improvement over the work of Zhang and Polymath.

Firstly, improving on the Polymath bound of 4680, Maynard showed that there are infinitely many pairs of primes that differ by at most 600. Let me repeat that: just a few months after Zhang made the breakthrough by obtaining the first finite bound of 70 000 000, Maynard brought that all the way down to 600!

Secondly, Maynard and Tao were able to say something about gaps between other pairs of primes, not just consecutive primes, thereby guaranteeing the existence of little clusters of primes nestled close together. Another way of phrasing the Twin Primes Conjecture would be to say that there are infinitely many intervals of length 2 that contain at least 2 primes (in fact it must be exactly 2 primes). For example, the interval from 17 to 19, including both endpoints, is an interval of length 2 that includes 2 primes. There's some standard notation that's relevant here: for each number m, we let H_m be the smallest number (if one exists) with the property that there are infinitely many intervals of length H_m that contain at least $m + 1$ primes. For example, the Twin Primes Conjecture predicts that $H_1 = 2$, and Polymath proved that $H_1 \leq 4680$. Knowing that H_m is finite guarantees that there are infinitely many clusters of $m + 1$ primes that are close together.

Remarkably, Maynard and Tao were able to show for the first time that for each number m, the number H_m is finite. Maynard's bound was slightly better than Tao's, and is a formula depending on m, but for technical reasons it's not very convenient to write down an actual number. This was a breakthrough in the same sort of vein as Zhang's: the precise number obtained is unimportant, what matters is that we know that there *is* a finite number that works. For example, there's some finite number with the property that there are infinitely many intervals of that length that contain at least 3 primes. When you think about it, that's pretty impressive stuff: not just two primes close together, but three, or even more—although of course the length of the intervals increases as we look for more primes.

Their ideas involved bypassing the more complicated approach of Zhang, and instead returning to the work of Goldston, Pintz and Yıldırım. Maynard and Tao both realised that they could adapt these ideas to prove bounded gaps between primes *without* relying on the unproven Elliott–Halberstam Conjecture that Goldston, Pintz and Yıldırım had needed. Zhang managed to improve on the work of Goldston, Pintz and Yıldırım by both showing that he could rely on a weaker statement about the distribution of the primes than the Elliott–Halberstam Conjecture, and then going on to prove that weaker statement. Maynard and Tao bypassed all of that, going back to the original sieve of Goldston, Pintz and Yıldırım, and adapting it to a more flexible sieve that yielded good bounds directly, relying only on the Bombieri–Vinogradov Theorem (a weaker form of the Elliott–Halberstam Conjecture that has, crucially, been proven, as we saw in Chapter 10). Moreover, using this approach Maynard and Tao were able to improve on the bounds from Zhang and Polymath. Tao obtained a slightly weaker bound than Maynard, but the approaches were similar.

We saw back in Chapter 7 that Zhang showed that for a suitable quantity k_0, given an admissible set of size k_0 (punch-card with k_0 holes) there are infinitely many sets of visible numbers that contain at least two primes. Zhang was able to take $k_0 = 3\,500\,000$, and then Polymath incrementally brought this down to 632. Maynard's approach allowed him to take $k_0 = 105$. By this point there was an online library of admissible sets, maintained by the MIT mathematician Andrew Sutherland. Thomas Engelsma submitted an admissible set of size 105 on 27th June 2013 (he was something of an expert, having experimented with finding explicit examples of admissible sets years before the Polymath project!), and it was this that Maynard used in his paper. The set is

$$\{0, 10, 12, 24, 28, 30, 34, 42, 48, 52, 54, 64, 70, 72, 78, 82, 90, 94, 100, 112,$$
$$114, 118, 120, 124, 132, 138, 148, 154, 168, 174, 178, 180, 184, 190, 192, 202,$$
$$204, 208, 220, 222, 232, 234, 250, 252, 258, 262, 264, 268, 280, 288, 294, 300,$$
$$310, 322, 324, 328, 330, 334, 342, 352, 358, 360, 364, 372, 378, 384, 390, 394,$$
$$400, 402, 408, 412, 418, 420, 430, 432, 442, 444, 450, 454, 462, 468, 472, 478,$$
$$484, 490, 492, 498, 504, 510, 528, 532, 534, 538, 544, 558, 562, 570, 574, 580,$$
$$582, 588, 594, 598, 600\}.$$

I have included that simply to make the point that I can: it's a small enough set, involving small enough numbers, that I can copy it without risk of upsetting my publisher! That gives just some indication of the dramatic improvements on Zhang's original bounds (where I really, really wouldn't want to type out an acceptable admissible set). For Maynard's bound, the crucial thing here is that the set has width 600. If we made a punch-card from this admissible set, Maynard's theorem would tell us that there are infinitely many positions of this punch-card where at least two of the visible numbers are prime—and these two numbers must differ by at most 600, because that's the width of the set.

Better still, as I mentioned, the new sieve approach of Maynard and Tao yields more than this, because it tells us about finding more than two primes in a short interval. Their sieve argument requires the use of a fact about the distribution of the primes. They were able to rely only on the Bombieri–Vinogradov Theorem, which says that the primes have 'level of distribution θ' for any $\theta < 1/2$, and Maynard was able to squeeze out the bound of 600 for pairs of primes from this. If, however, we allow ourselves to assume more (that is, more than we currently know how to prove), then the bounds are correspondingly better. Assuming the Elliott–Halberstam Conjecture, that the primes have level of distribution θ for any $\theta < 1$, Maynard can show that there are infinitely many pairs of primes that differ by at most 12, and also that there are infinitely many triples of primes where the first and third differ by at most 600. Wow! Let me say that again. Assuming the Elliott–Halberstam Conjecture, Maynard's work shows that there are infinitely many pairs of primes that differ by at most 12.

Maynard had shown great potential as a graduate student, and this work in his first year as a postdoc was fulfilling that potential in an impressive way. Since his year in Montreal, Maynard has become a 'Fellow by Examination' (Junior Research Fellow) at Magdalen College, Oxford, has been awarded the 2014 SASTRA Ramanujan Prize 'for outstanding contributions by young mathematicians to areas influenced by the genius Srinivasa Ramanujan', has been awarded a 2015 Whitehead Prize by the London Mathematical Society, and from July 2015 Maynard has been a Clay Research Fellow. The mathematical community has been quick to recognise the significance of Maynard's work, not only because of the results he obtained, but also because of the ideas that he has introduced that have the potential to lead to further breakthroughs.

Spurred on by the new ideas from Maynard and Tao, the Polymathematicians who had worked on bounded gaps between primes resurrected the project, looking to see how much more progress might be made building on the recent leaps forward. The gap was closing again ...

CHAPTER 14

Generalise . . .

We have spent quite some time thinking about which numbers can be written as a sum of two squares and which can't. Having answered that question, it seems natural to consider some other related questions. Can we generalise our work?

Sums of more squares

Here's one direction we could take. We've thought about writing numbers as sums of two squares. What if we allow ourselves *three* squares? That's going to expand the pool of possible numbers, but can we say anything about which numbers we obtain? For example, 6 is not a sum of two squares but is a sum of three squares ($6 = 2^2 + 1^2 + 1^2$). Since I'm including 0 as a square number, anything that's a sum of two squares is also a sum of three squares, so we really are *expanding* the pool. Let's have a look on the 8×8 grid (Figure 14.1). Here the blue numbers are sums of two squares, and the yellow are the extra numbers that are sums of three squares but not of two squares.

Once again, the grid is strongly suggesting some interesting patterns. Perhaps the first thing that strikes me is that by allowing ourselves a third square, we've been able to reach a lot more numbers—there are a lot of yellow numbers on that grid. Nonetheless, there are a whole bunch of numbers that aren't a sum of three squares.

The 7 column is looking pretty uncoloured: none of the numbers up to 100 that are 7 more than a multiple of 8 can be written as the sum of three squares. Will that pattern continue?

1	2	3	4	5	6	7	8
9	10	11	12	13	14	15	16
17	18	19	20	21	22	23	24
25	26	27	28	29	30	31	32
33	34	35	36	37	38	39	40
41	42	43	44	45	46	47	48
49	50	51	52	53	54	55	56
57	58	59	60	61	62	63	64
65	66	67	68	69	70	71	72
73	74	75	76	77	78	79	80
81	82	83	84	85	86	87	88
89	90	91	92	93	94	95	96
97	98	99	100				

Fig. 14.1 Blue sums of two squares, and yellow sums of three squares that are not sums of two squares.

Happily, we already have a strategy for thinking about this. We know (from Chapter 12) that every square is a multiple of 8, or 1 more than a multiple of 8, or 4 more than a multiple of 8. So what can we get by adding three of those, and, more to the point, what can we *not* get?

If we think about it a bit, we find that we can reach each remainder except 7: there is no way to obtain 7 by adding three numbers that are each 0, 1 and 4. And so the 7 column in our grid is genuinely empty, a number that is 7 more than a multiple of 8 is never a sum of three squares.

Modular arithmetic

This kind of argument, thinking about remainders, is proving really helpful, we've used it a few times now. It's so useful that mathematicians have invented language and notation for it, because writing things like 'is 1 more than a multiple of 8' becomes a bit inconvenient when you're doing it frequently. This kind of thinking is called *modular arithmetic*. Let me just briefly describe the notation, which captures the ideas we've been discussing already. Here are six ways of saying the same thing.

- 123 and 51 leave the same remainder on division by 8.
- 123 is 51 more than a multiple of 8.

- $123 = 51 + 8n$ for some integer n.
- $123 - 51$ is divisible by 8.
- 8 divides $123 - 51$.
- 123 is congruent to 51 modulo 8, written $123 \equiv 51 \pmod{8}$.

The symbol $\equiv$ is like an equals sign $=$ but with three horizontal bars instead of two. It means 'is congruent to', so for example I would read $123 \equiv 51 \pmod{8}$ out loud as '123 is congruent to 51 modulo 8'.

It might seem a bit laboured to spell out so many ways of saying the same thing, but in mathematics it is often really helpful to be able to switch between multiple ways: sometimes the key to making progress on a problem can be to adopt a different perspective.

Often when working modulo n we want to focus on remainders on division by n, so we think of each number as being congruent to one of $0, 1, 2, \ldots, n - 1$, but it's worth being aware that there's no rule about this, that's why I used 51 as an example above. Students sometimes fall into the bad habit of thinking that the number on the right-hand side of the congruence symbol $\equiv$ must always be less than the modulus, but this is simply not the case. In addition to the statements above, it would be true to say $123 \equiv 3 \pmod{8}$ and $51 \equiv 3 \pmod{8}$, and also $123 \equiv -5 \pmod{8}$ and $123 \equiv -37 \pmod{8}$.

A small piece of algebra (that I'm not going to do here, but that you might like to think about) shows that addition, subtraction and multiplication all behave nicely in modular arithmetic: if $a \equiv b \pmod{n}$ and $c \equiv d \pmod{n}$, then $a + c \equiv b + d \pmod{n}$ and $a - c \equiv b - d \pmod{n}$ and $ac \equiv bd \pmod{n}$. For example, we know that $123 \equiv 3 \pmod{8}$, and so we know that $123^2 \equiv 3^2 \equiv 1 \pmod{8}$—crucially, I did *not* have to work out 123^2 in order to deduce that! But beware, division in modular arithmetic is an extremely dangerous business, and should be approached with extreme caution. (For a start, modular arithmetic is about integers, and it's entirely possible that dividing one whole number by another does not lead to a whole number.) There's lots of interesting stuff I could say here about division in modular arithmetic, but we don't need it right now so I shall resist the temptation to take a detour.

Back to sums of three squares

Let's see how our sums-of-three-squares argument looks with the notation of modular arithmetic.

Modulo 8, every number is congruent to exactly one of 0, 1, 2, 3, 4, 5, 6 or 7 (just the remainder on division by 8). In order to compute the squares modulo 8, we can simply square each of those 8 numbers. We have

$$\begin{aligned}
0^2 &\equiv 0 \pmod{8},\\
1^2 &\equiv 1 \pmod{8},\\
2^2 &\equiv 4 \pmod{8},\\
3^2 &\equiv 1 \pmod{8},\\
4^2 &\equiv 0 \pmod{8},\\
5^2 &\equiv 1 \pmod{8},\\
6^2 &\equiv 4 \pmod{8}, \text{ and}\\
7^2 &\equiv 1 \pmod{8},
\end{aligned}$$

So the squares modulo 8 are 0, 1 and 4. (Of course we already knew this, what we're doing here is establishing it using the notation of modular arithmetic.)

(Note that it's entirely possible for the squares modulo something *not* to be normal squares. For example, 2 is a square modulo 7, because $3^2 \equiv 2 \pmod{7}$. It just so happens that modulo 8 the squares are all squares in the real world too.)

Now we need to consider the sums of three squares modulo 8. There is no way to obtain 7 by adding three numbers that are each 0, 1 or 4, so there is no solution to the congruence $a + b + c \equiv 7 \pmod{8}$ where a, b and c are 0, 1 or 4, so there are no integers x, y and z with $x^2 + y^2 + z^2 \equiv 7 \pmod{8}$. This means that if a number is congruent to 7 modulo 8 then it cannot be written as a sum of three squares. (Note that this does not show that if a number is not congruent to 7 modulo 8 then it *can* be written as the sum of three squares, it just rules out all the numbers that are congruent to 7 modulo 8.)

This really is exactly the argument we saw before, just written in a slightly more compact way. The power of modular arithmetic becomes more apparent with more complex and intricate arguments, and it is a vital tool from university level onwards.

Let's go back to the grid. The 7 column is empty, we've established that. What else? There are a few other gaps, what can we say about them? Well, they're 28, 60 and 92. If you look carefully, you might notice that they're 4 times the first three numbers in the 7 column. One of my guiding principles when thinking about mathematics is that *there are no coincidences*. It simply cannot be a coincidence that they're related to the 7 column in that way: it must be a sign that there's something interesting going on.

For sums of two squares, we were able to find a beautiful theorem classifying which numbers are of that form. It turns out to be possible to do the same for sums of three squares, although proving it is harder (so I'm not even going to try to do it here).

Theorem *A positive integer is a sum of three squares if, and only if, it is* not *of the form* $4^a(8b+7)$ *for integers a and b.*

Notice that the bad numbers here, the numbers that are not sums of three squares, are precisely the numbers in the 7 column, 4 times the numbers in the 7 column, 4^2 times the numbers in the 7 column, and more generally any power of 4 times the numbers in the 7 column. Importantly, this reveals that our prediction based on looking at the numbers up to 100 is not quite right. For example, 240 is of this form (since $240 = 4^2 \times 15$ and 15 is in the 7 column) and so cannot be written as a sum of three squares, but 240 itself is in the right-hand 8 column and so we might have guessed that it would be a sum of three squares.

It's not too hard to show that numbers of this form (power of 4 times number in 7 column) cannot be written as a sum of three squares, using an argument modulo 8 such as above (indeed, we've already dealt with some of these numbers, so it's a case of extending that argument to the more general case). The difficult part is showing that all the remaining numbers really *can* be written as a sum of three squares, and that's the bit I'm not going to discuss further here.

Sums of four squares

Instead, let's press on. Sums of two squares and sums of three squares have been interesting, so what about sums of four squares? This is shown in Figure 14.2, where this time the blue numbers are sums of three squares and yellow are the extra numbers that are sums of four squares but not of three.

Wow! We've been able to fill in all those gaps: every number up to 100 can be written as a sum of four squares. That's clearly a consequence of our working with small numbers, though, surely it can't be the case for *all* numbers, can it? For our grid, with hindsight it's not such a surprise, because all of the gaps in the grid for three squares are just one more than a number that *is* a sum of three

1	2	3	4	5	6	7	8
9	10	11	12	13	14	15	16
17	18	19	20	21	22	23	24
25	26	27	28	29	30	31	32
33	34	35	36	37	38	39	40
41	42	43	44	45	46	47	48
49	50	51	52	53	54	55	56
57	58	59	60	61	62	63	64
65	66	67	68	69	70	71	72
73	74	75	76	77	78	79	80
81	82	83	84	85	86	87	88
89	90	91	92	93	94	95	96
97	98	99	100				

Fig. 14.2 Blue sums of three squares, yellow sums of four squares that are not sums of three squares.

squares, so we can reach the number in the gap by adding 1^2 to a sum of three squares. But will that continue?

Well, ... astonishingly it turns out that this is not just a quirk of small numbers. This is a really surprising theorem of Joseph-Louis Lagrange (1736–1813), whom we met briefly in Chapter 4 as a correspondent and supporter of Sophie Germain.

Theorem (Lagrange) *Every positive integer can be written as a sum of four squares.*

That's it. *Every* positive integer. No exceptions. I first came across this theorem quite a few years ago now, and I still feel amazed that it's true!

It turns out that having proved the theorem about sums of three squares, it's not too hard to deduce the result about sums of four squares. But interestingly the latter is actually easier to prove than the theorem of sums of three squares, so if you just want to know Lagrange's theorem then it's easiest to prove it directly.

One key reason for this is inspired by sums of two squares. You might remember that an important part of our argument about sums of two squares was that if we take two numbers that are sums of two squares and multiply them, then we get another number that is a sum of two squares. Rather satisfyingly, the same is true for sums of four squares, and this is a useful step in the argument (it means that we can focus our attention on showing that every prime is a sum of four squares, and that will finish the argument). But it's just plain false for sums of three squares. We can go back to our grid to find a counterexample: we know that $3 = 1^2 + 1^2 + 1^2$ and $5 = 2^2 + 1^2 + 0^2$ are sums of three squares, but that $3 \times 5 = 15$ is not.

What makes it work for four squares? Well, as for sums of two squares there's a short but unsatisfying algebraic justification, and a deeper reason behind it. The algebraic identity this time is that

$$(a^2 + b^2 + c^2 + d^2)(x^2 + y^2 + z^2 + w^2)$$

is equal to

$$(ax-by-cz-dw)^2+(bx+ay-dz+cw)^2+(cx+dy+az-bw)^2+(dx-cy+bz+aw)^2$$

for all a, b, c, d, x, y, z and w.

True, but curiously uninformative. (Mathematics isn't just about equations, it's also about the reasoning behind the equations!) For sums of two squares, the underlying explanation turned out to be informed by complex numbers. This time, we want a generalisation of the complex numbers called the *quaternions*, first studied by the Irish mathematician William Rowan Hamilton (1805–1865) in the middle of the nineteenth century. The story is that the idea came to him in a flash while he was walking through Dublin, and he was so inspired that he carved the description of quaternions into the stone bridge he was crossing. A rare example of mathematical graffiti!

A complex number has the form $a + bi$, where a and b are real numbers and i has the property that $i^2 = -1$.

A quaternion has the form $a+bi+cj+dk$, where a, b, c and d are real numbers, and i, j and k have the properties that $i^2 = j^2 = k^2 = ijk = -1$. The secret here is not to worry about what i, j and k *are*, but just to focus on what they *do*—and what they do is governed by $i^2 = j^2 = k^2 = ijk = -1$.

For example, what can we say about ij? Well, if we take $ijk = -1$ and multiply by k, we get $ijk^2 = -k$. But $k^2 = -1$, so that becomes $-ij = -k$, so $ij = k$.

The quaternions come with a health warning. They do not behave in all respects as we might hope! In particular, having found that $ij = k$, let's see what we can say about ji.

Well, $j^2 = -1$, so $ji = -jij^2 = -j(ij)j = -jkj$ using our work on ij.

And $ijk = -1$ so if we multiply both sides by i then we get $i^2 jk = -i$, so $jk = i$ (using $i^2 = -1$). Using that in the expression from the previous paragraph gives $ji = -(jk)j = -ij = -k$.

Strange, isn't it?! The order in which we carry out the multiplication matters. We know that $789 \times 456 = 456 \times 789$ even without bothering to do the multiplication. In the jargon, *multiplication is commutative*. (I love saying that, it makes me think of the Tom Lehrer song *New Math*—although he's talking about addition rather than multiplication.) But in the quaternions, this familiar, comforting property no longer holds. The world has turned upside down, and we have to concentrate very, very hard. In particular, when I said 'multiply by k', I should really have said 'multiply on the right by k' (meaning do blah$\times k$, not $k\times$blah), and when I said 'multiply both sides by i' I should have said 'multiply both sides on the left by i').

So the quaternions are cool, exciting, and a bit weird. Why are they relevant to sums of four squares? Well, in the same way that to a complex number $a + bi$ we assign a quantity $a^2 + b^2$ (and think of it as the square of its length), to a quaternion $a + bi + cj + dk$ we assign the quantity $a^2 + b^2 + c^2 + d^2$. Then the identity on the previous page tells us what happens when we multiply these. So, by analogy with complex numbers (this is such a great way of doing mathematics, by the way: you take what you did before and see what you can copy with suitable modifications for the next problem), I generated the above identity by taking two quaternions, multiplying them, and using that to give the product as a sum of four squares. Let me show you what I mean. We take two quaternions, say $a + bi + cj + dk$ and $x + yi + zj + wk$, and multiply them. As we go, we can simplify, using the rules $i^2 = j^2 = k^2 = ijk = -1$ and also the rules we obtained above: $ij = k$, $ji = -k$, and similarly $jk = i$, $kj = -i$, $ki = j$ and $ik = -j$. We find (go on, do the multiplication yourself, don't just believe me!) that the product is

$$(ax - by - cz - dw) + (bx + ay - dz + cw)i + (cx + dy + az - bw)j + (dx - cy + bz + aw)k.$$

And *that* is what inspired the identity above. I'll leave you to multiply out both sides of my claimed equation about sums of four squares to check that it really is true ...

Once we have proved this result, it is enough to show that every prime number is a sum of four squares. That's because if we take any whole number bigger than 1, we can consider its prime factorisation, use the fact that every prime is a sum of four squares, and then use the fact that if we multiply sums of four squares then we get another sum of four squares. Proving that every prime is a sum of four squares takes some work, and I'm not going to go into the details here, but let's just look at an example to see why this, combined with the argument above about multiplying sums of four squares, shows that *every* whole number greater than 1 is a sum of four squares.

For example, let's write 247 as a sum of four squares. That number is small enough that we could just do it by experimentation, but that wouldn't be in the spirit of illustrating the ideas above!

The prime factorisation of 247 is 13×19, so let's focus on those primes. We have $13 = 3^2 + 2^2 + 0^2 + 0^2$ and $19 = 4^2 + 1^2 + 1^2 + 1^2$. Using our previous identity that shows that the product of two sums of four squares is another sum of four squares, we see that

$$\begin{aligned}(3^2 + 2^2 + 0^2 + 0^2)(4^2 + 1^2 + 1^2 + 1^2) &= (12 - 2)^2 + (8 + 3)^2 + (3 - 2)^2 + (2 + 3)^2 \\ &= 10^2 + 11^2 + 1^2 + 5^2,\end{aligned}$$

—and indeed $247 = 10^2 + 11^2 + 1^2 + 5^2$.

Quaternions are an intriguing mathematical curiosity, and were good for generating a useful identity about sums of four squares, but they are not just a toy: they turn out to be enormously helpful in various applications (as well as in other branches of mathematics). In particular, they give a very good way of describing rotations, and consequently are used in physics and also in computer graphics, for example.

A different sort of generalisation

So far in this chapter our strategy has been to increase the numbers of squares we're allowed to add: we started with sums of two squares, then sums of three squares and then sums of four squares. But Lagrange's theorem that every positive integer is a sum of four squares means that this game is now at an end. The question of which numbers are sums of five squares is extremely boring, because the answer is clearly 'all of them'.

So where next? Well, squares were good, but what about higher powers? (This is like Fermat's response to Diophantus's writing about integer solutions to the Pythagorean equation $a^2 + b^2 = c^2$: what if we replace the squares by higher powers?) The problem I want to concentrate on here is trying to generalise Lagrange's theorem. We know that every positive integer is a sum of four squares. Is a similar result true for cubes? Presumably that result would be of the form 'every positive integer is a sum of ______ cubes', where we have to think of a suitable number to fill in the blank. And what about powers beyond cubes?

These questions were considered by the Cambridge mathematician Edward Waring (1736–1798) in the eighteenth century. Waring worked on various topics in algebra and number theory, and had quite significant ideas, although it seems that his writing was rather obscure and so other mathematicians perhaps paid his work less attention than it deserved. As an aside, let me mention a nice link between Waring and our story of prime numbers: Waring also proposed Goldbach's Conjecture (as it were). Although Goldbach made his famous conjecture (about even numbers being sums of two primes) years before Waring, he did so in a letter to Euler, so when Waring proposed the conjecture in a book he was the first to publish the claim.

Anyway, the reason that Waring is best known today is for *Waring's problem*, in which he built on Lagrange's theorem to seek to generalise to higher powers. He claimed that

> Every integer is equal to the sum of not more than 9 cubes. Also every integer is the sum of not more than 19 fourth powers, and so on ....

—but he offered no proof, and so demonstrating the truth of this statement became known as Waring's problem.

Before anyone can think about proving Waring's conjecture, though, we need to make a bit more sense of it. He claims that every (positive) integer is the sum of nine cubes. (He says 'not more than', but we can omit that because we're including $0^3 = 0$ as a cube. For example, $5 = 1^3 + 1^3 + 1^3 + 1^3 + 1^3$ is a sum of five cubes, and hence of nine with our interpretation.) That is a clear and precise conjecture. Similarly, he claims that every (positive) integer is the sum of nineteen fourth powers. Again, so far so good. But then he says 'and so on', which is a bit vague! What does Waring mean by this?

The way in which we interpret this is as follows. Pick a power to concentrate on, say k^{th} powers. (If you prefer numbers to letters, then you could replace k^{th} by say 7th throughout this.) Then there is some number s (which is allowed to depend on k) so that every positive integer is a sum of s k^{th} powers.

For example, Lagrange's theorem says that for squares we can take $s = 4$: every positive integer is a sum of four squares. Waring explicitly conjectured that for cubes we can take $s = 9$, and for fourth powers we can take $s = 19$, but more generally the idea is that we can always find *some* number that works.

Waring's conjecture did not predict a rule for finding such numbers, just that there always will be a number.

As so often in the course of number theory, this turns out to be a rather hard problem—but this one has been solved. In 1909, the distinguished German mathematician David Hilbert (1862–1943) gave a solution: he proved that Waring's conjecture is true, for any k there is an s (depending on k) so that every positive integer is a sum of s k^{th} powers. So that was the end of that.

Well, not quite. This is one of those instances in mathematics where the first proof was important for demonstrating the truth of the statement, but where a subsequent argument turned out to be at least as important. In the 1920s, G.H. Hardy and J.E. Littlewood (whom we met briefly in Chapter 6) developed a whole new approach to problems such as this, now known as the Hardy–Littlewood circle method. The circle method is important both because it offers new insights into Waring's problem and because mathematicians have been able to apply it to other problems too.

The Hardy–Littlewood circle method

One of the intriguing features of the Hardy–Littlewood circle method is the counterintuitive key to its success, which is that Waring's problem becomes easier to solve if we instead try to solve a harder problem. This is a phenomenon within mathematics that I find fascinating—sometimes trying to solve a harder problem somehow can make it easier to make progress. How does that work for Waring's problem? Well, we take a fixed k and our goal is to show that there is some s such that every positive integer is a sum of s k^{th} powers. Hardy and Littlewood set out to do more than that. They decided not just to show that (for suitable s) each number is a sum of s k^{th} powers, but to count the number of ways to write each number as a sum of s k^{th} powers. Then 'all' we have to do is to show that the number of such ways is always at least 1!

Hardy and Littlewood didn't get an exact formula for the number of representations, that would be overly ambitious, but they did obtain an *asymptotic formula*. This is an approximate formula, with two important features. One is that while it is an approximation, it is an approximation with an error term, so we have a very precise measure of how far from reality the formula can be. The other is that it becomes increasingly accurate as we're thinking about larger numbers. If you want to know how many ways there are to write a thousand as a sum of kth powers, then get a computer to find all the possibilities and count them. But if you want to know how many ways there are to write some absolutely massive number as a sum of kth powers, then the Hardy–Littlewood asymptotic formula will give you a pretty good idea.

The asymptotic formula

I think that I'll tell you what the asymptotic formula is, just for fun, and then I'll unpick it a bit (so don't be intimidated by it!).

Theorem (Hardy–Littlewood) *Fix* $k \geq 2$. *If* $s \geq 2^k + 1$, *then the number of solutions to the equation* $N = x_1^k + x_2^k + \cdots + x_s^k$ *is*

$$\frac{\Gamma(1+\frac{1}{k})^s}{\Gamma(\frac{s}{k})}\mathfrak{S}(N)N^{\frac{s}{k}-1} + O(N^{\frac{s}{k}-1-\delta})$$

for some fixed $\delta > 0$, *where*

$$\mathfrak{S}(N) = \sum_{q=1}^{\infty} \sum_{\substack{a=1 \\ (a,q)=1}}^{q} \left(\frac{1}{q}\sum_{n=1}^{q} e^{2\pi i \frac{an^k}{q}}\right)^s e^{-2\pi i \frac{Na}{q}}$$

and $\mathfrak{S}(N) \geq C(k,s) > 0$.

Right. What does that actually mean?! Let's focus on the asymptotic formula. We think of this as having two terms, the *main term* and the *error term*:

$$\underbrace{\frac{\Gamma(1+\frac{1}{k})^s}{\Gamma(\frac{s}{k})}\mathfrak{S}(N)N^{\frac{s}{k}-1}}_{\text{main term}} + \underbrace{O(N^{\frac{s}{k}-1-\delta})}_{\text{error term}}.$$

Let's think about a couple of more friendly looking examples for a moment, and then we'll come back to this more complicated situation. Suppose we have a function of the form

$$f(N) = 0.000001N^3 - 1000000N.$$

How does this behave when N is really, *really* enormous? Well, the $0.000001N^3$ term is going to be much, much bigger than the $1000000N$ term. In fact, the $1000000N$ term is going to be negligibly small in comparison with $0.000001N^3$, even though the coefficient (the number in front) is much larger for the $1000000N$ term than for $0.000001N^3$. What matters is the exponent of N (3 versus 1), not the number out the front, because the coefficient is fixed and so can easily be dwarfed by the effect of the exponent when N is vast.

What if we think about a formula

$$g(N) = 0.000001N^3 - 1000000N^{2.99}?$$

Just the same thing happens. For enormous N, the first term $0.000001N^3$ is going to be much bigger than the second term $1000000N^{2.99}$, for just the same reason as before. Here the exponents are much closer, so we're going to need much bigger N for the effect to become apparent, but we don't care about that!

In both of the above examples, the form of the expression makes it clear that if N is large enough then the expression will be positive: if N is big enough then $f(N) > 0$, and if N is big enough then $g(N) > 0$ (where 'big enough' might mean different things in those two cases, but I don't mind that at all).

In our case, we have a formula that is *counting* something. In particular, we know that the number of solutions to the equation (the number of representations of N as a sum of s k^{th} powers) must be a whole number. Here's one of my favourite little observations in mathematics: the smallest positive integer is 1. So if we have a number that we know is a whole number, and that we know is positive, then we know it must be at least 1. Remember that our goal is to show that the number of solutions to the equation is at least 1, so it's enough to show that it's strictly positive. Our examples above have shown us the way forwards: we need to know that the main term 'beats' the error term if N is large enough.

Why do we not need to worry about the 'if N is large enough'? Well, we only really care about showing that every large enough N is a sum of s k^{th} powers, because in principle that leaves a finite problem (albeit potentially such a vast one that even with all the computers in the world we wouldn't be able to check it in the lifetime of the universe, or something). Mathematically, the most interesting part is the problem for N beyond some fixed point.

Now back to our main term and error term. Let me tell you a bit more about the main term first. There are some very strange symbols here. Right at the front, Γ is the Gamma function. It doesn't matter too much here what exactly that is, all we need to know is that

$$\frac{\Gamma(1+\frac{1}{k})^s}{\Gamma(\frac{s}{k})}$$

is a positive number, and *it doesn't depend on N*. It does depend on s and on k, but we're thinking of them as fixed. For our purposes, it's a complicated way of describing a fixed positive number. Then there's $\mathfrak{S}(N)$, which is 'helpfully' defined in the statement of the theorem. I'll come back to what it tells us in a moment. It's called the *singular series*, by the way (the symbol $\mathfrak{S}$ is a rather flamboyant 'S'). What we need to know now is that although it really does depend on N, it's always at least as big as some positive constant $C(k, s)$ (as asserted in the theorem). I say 'constant' because although $C(k, s)$ can depend on k and on s, we're thinking of those as fixed. The important point is that no matter what N is, we know that $\mathfrak{S}(N)$ is always 0.0000001 or something (the actual value is unimportant).

What this allows us to deduce is that the main term is always greater than or equal to some positive constant times $N^{\frac{s}{k}-1}$. This is analogous to the $0.000001N^3$ that we considered in our examples above. As long as the error term is less than a smaller power of N (even if it's only a slightly smaller power), we'll know that for large enough N the main term beats the error term, and so that the number of representations is positive (and hence at least 1) for large enough N.

The error term uses some technical notation too. It's $O(N^{\frac{s}{k}-1-\delta})$. Here, the $O()$ means that the whole term grows at most as fast as a constant times the stuff inside the brackets. So the error term is always between $-1000000N^{\frac{s}{k}-1-\delta}$ and $1\,000\,000N^{\frac{s}{k}-1-\delta}$, where I've made up that number 1 000 000, it might need to be a much larger value. There's a slightly curious kind of thought process needed for these arguments, where we simultaneously need to make extremely careful and precise statements (because we're mathematicians), and at the same time we need to focus on the important points and manage to filter out the unimportant ones. We saw in the examples above that the coefficients, the numbers out the front, aren't important. What matters are the powers of N, and so the *big O notation* (yes, that really is what it's called) helps to strip out the irrelevant coefficients so that we focus on what matters.

The theorem includes a little comment 'for some fixed $\delta > 0$'. So what we know is that there's some positive number (potentially extremely tiny) so that the error term behaves like a number times $N^{\frac{s}{k}-1-\delta}$. Now the significance of the main term and the error term starts to become clear. The main term is like N to the power $\frac{s}{k} - 1$, and the error term is like N to a power that is just a little bit smaller. It's very similar to our example from earlier where the main term was like N^3 and the error term was like $N^{2.99}$.

We've seen that for (very) large N the error term will be insignificant in comparison to the main term, and so we can deduce that the number of solutions to the equation (which is what this asymptotic formula is counting) will be positive and hence at least 1. Bingo!

The singular series

I said that one of the strengths of the Hardy–Littlewood circle method for Waring's problem is that it gives us more insight into what's really going on. That it gives us an asymptotic formula for the number of representations of a large number as a sum of s k^{th} powers already ticks that box, but what is particularly satisfying about the formula is that its various components come from sensible places. The main term is not just a mysterious string of symbols: each of the pieces sheds some light on the problem.

I'd like to try to give a sense of what the singular series $\mathfrak{S}(N)$ tells us, because I think it's rather nice. The statement of the theorem defines it as

$$\mathfrak{S}(N) = \sum_{q=1}^{\infty} \sum_{\substack{a=1 \\ (a,q)=1}}^{q} \left(\frac{1}{q} \sum_{n=1}^{q} e^{2\pi i \frac{an^k}{q}} \right)^s e^{-2\pi i \frac{Na}{q}}.$$

This is frankly terrifying, but also rather puzzling. We're trying to count the integer solutions to an equation. How can the answer possibly involve infinite sums, the exponential function, the number π and even i, the square root of -1?! This is one of the things I love about the Hardy–Littlewood circle method: although we're tackling a question in number theory—we're studying whole numbers—nonetheless we need to draw on all sorts of areas of mathematics in order to make sense of the problem. Mathematics is interconnected: when you embark on a question, you never quite know what tools and techniques and ideas you might want to use.

Let's go back to squares for a bit, and let's work modulo 8 again as that was what we did earlier (that is, we're thinking about remainders on division by 8). We saw previously that every square is a multiple of 8 or 1 more than a multiple of 8 or 4 more than a multiple of 8. That is, every square is congruent to 0 or 1 or 4 modulo 8. Still working modulo 8, how many ways are there to write 0 as a sum of four squares? Well, we have

$$0 \equiv 0+0+0+0 \pmod 8,$$
$$0 \equiv 0+0+4+4 \pmod 8,$$
$$0 \equiv 4+4+4+4 \pmod 8$$

and a bunch of rearrangements of those (such as $0 \equiv 0 + 4 + 0 + 4 \pmod 8$). With a bit of thought, we can see that there's no way to use 1 if we're hoping to write 0 as a sum of four squares, and so the solutions above are the only ones.

What if we now try to write 7 as a sum of four squares (again working modulo 8)? This time there's

$$7 \equiv 1 + 1 + 1 + 4 \pmod 8$$

but there are no other possibilities (apart from reordering the numbers).

We see that modulo 8, 0 and 7 can be written as sums of four squares in different numbers of ways.

This information really ought to have an effect on the number of representations of numbers as sums of actual squares (not working modulo anything)—we don't expect the same number of representations for a multiple of 8 and for a number that is 7 more than a multiple of 8.

At the same time, there are going to be lots of other similar pieces of information obtained by working modulo each different number in turn, and this is what the singular series encodes. The singular series starts with a sum over q: for each value of q in turn we compute a quantity that records how easy or hard it is to represent N as a sum of s k^{th} powers when working modulo q.

The rest of the main term, $\frac{\Gamma(1+\frac{1}{k})^s}{\Gamma(\frac{s}{k})}$, also has a natural explanation, this time arising from thinking about how easy or hard it is to solve the equation $N = x_1^k + \cdots + x_s^k$ where $x_1, \ldots, x_s$ can be any positive reals (not just whole numbers).

Having given it some thought, I have decided not to try to describe the process by which the asymptotic formula is proved. I know that this will frustrate some readers, but the reason is maybe interesting in itself. The Hardy–Littlewood circle method relies on calculus in an entirely fundamental way (and I'm trying to write a book that doesn't require knowledge of calculus). Calculus! To solve a problem about whole numbers. I find it both unexpected and intriguing that calculus is such a useful tool for problems like this—but it really is. It turns out that we can represent the number of integer solutions to the equation using an integral (from a modern perspective we obtain the integral via Fourier analysis), and then (with quite a lot of work, some quite fiddly estimates and some ingenuity) we can approximate the integral to obtain the asymptotic formula. However, this is not a place for a detailed discussion of that, so I'll say no more.

Ongoing work on Waring's problem

I would like to say just a bit more about Waring's problem, though, because in some sense the problem is still not solved. Lagrange's theorem tells us that every positive integer is a sum of four squares, and we know that this is 'best possible': we cannot replace 'four' by 'three' here because there are numbers that are not a sum of three squares (for example, 7). We know that for k^{th} powers there is *some* number s so that every number is a sum of s k^{th} powers, but what is the best possible value?

Waring asserted that every number is a sum of nine cubes. The theorem of Hardy and Littlewood shows that this is true for every sufficiently large number. This highlights that there are really two questions one might ask here. One is 'What is the smallest s so that *every* positive integer is a sum of s k^{th} powers?', and the answer to this question is conventionally called $g(k)$. The other is 'What is the smallest s so that *every sufficiently large* positive integer is a sum of s k^{th} powers?', and this is denoted by $G(k)$. The second question is arguably more interesting than the first, because the answer to the first can be skewed a lot by the effects of small numbers. For example, what if we want to write 31 as a sum of 5th powers? Since $2^5 = 32$ is too big, the only fifth power we have available to us is $1^5 = 1$, so we need 31 powers in order to write 31. But this is a somewhat contrived example: I deliberately picked $2^5 - 1$ in order to give a number that needs a lot of fifth powers. Perhaps for very large numbers we can get away with fewer powers, because there are more fifth powers available to us.

We know that for squares we have $g(2) = 4$, so $G(2) \leq 4$. In fact, $G(2) = 4$ too, because no number that is 7 more than a multiple of 8 can be written as a sum of three squares.

Hardy and Littlewood showed that $G(3) \leq 9$: we know that 9 cubes will suffice, perhaps a smaller number would work too. It turns out that $g(3) = 9$, because, for example, 23 is not a sum of eight cubes (but is a sum of nine cubes). Since Hardy and Littlewood, there have been improvements to their bound, and it is now known that $G(3) \leq 7$. But the exact value is not yet known—it is possible that $G(3)$ might be as small as 4.

More generally, the value of $g(k)$ is known for all $k \geq 2$. This is essentially because of the effect of small numbers that artificially push up the value (as with our example of 31 as a sum of fifth powers above). The more interesting question, of the value of $G(k)$, is known exactly only for two cases. One is Lagrange's theorem: as above, $G(2) = 4$. The other was proved by Harold

Davenport (1907–1969), who did his PhD in Cambridge under the supervision of Littlewood. In 1939 he showed that $G(4) = 16$ (this was just one of his many papers on Waring's problem). Incidentally, I like this line from Davenport:

> Mathematicians are extremely lucky, they are paid for doing what they would by nature have to do anyway.

Hardy and Littlewood's theorem gave upper bounds for $G(k)$ for general k, and throughout the twentieth century those bounds have been improved by successive number theorists. That work continues into the twenty-first century, as the quest to find exact values of $G(k)$ for all k remains unconcluded.

More on the Hardy–Littlewood circle method

The successes of the Hardy–Littlewood circle method are not confined to Waring's problem. The approach has turned out to be extremely powerful for tackling problems with the same flavour, roughly 'can we write every number of this type as a sum of some numbers of this type?' The way in which we tend to think about the method now owes a lot to Vinogradov, who developed and simplified Hardy and Littlewood's original approach in his work on the weak Goldbach Conjecture. As I mentioned in Chapter 6, Vinogradov was able to use this approach to show that every sufficiently large odd number is a sum of three primes.

I always feel slightly uncomfortable calling the Hardy–Littlewood circle method a method, because that slightly misrepresents the difficulties of actually implementing the method for a given problem. I think of it as more of a framework—it gives an outline structure for a proof, but there are then several substantial sections that need a lot of work to complete. This extends even to the Twin Primes Conjecture: it's possible to sketch out an outline argument using the Hardy–Littlewood framework. Unfortunately, there are some very serious technical obstacles to being able to complete the proof.

Let's return to the main quest of this book, by checking on the progress on understanding bounded gaps between primes.

CHAPTER 15

April 2014

Following Maynard's big leap forwards, in which he introduced new ideas that enabled him to prove that there are infinitely many pairs of primes that differ by at most 600, the Polymath8 project on 'Bounded gaps between primes' became Polymath8a, and Polymath8b was born. The goal of the new phase of the project, which started in November 2013, was to improve the bounds still further. There were a number of bounds under consideration. The key one from the point of view of the Twin Primes Conjecture was H_1, the smallest number so that there are infinitely many intervals of length H_1 that contain at least 2 primes. (The Twin Primes Conjecture predicts that $H_1 = 2$.) Maynard's work showed that H_1 is less than or equal to 600, so the challenge for the Polymathematicians was to improve on that, drawing on Maynard's own ideas, the arguments that were developed during Polymath8a, and anything else they could come up with! But, as we saw in Chapter 13, one of the breakthroughs of Maynard (and, at the same time, Tao) was, for the first time, to handle clusters of 3 or more primes, by giving a finite bound for H_m, the smallest number with the property that there are infinitely many intervals of length H_m that contain at least $m + 1$ primes. This bound demonstrates that not only are there infinitely many *pairs* of primes that are close, but also there are little clusters of three or four or even one hundred primes that are quite close together. Would the Polymathematicians be able to improve on the bounds that Maynard obtained?

Zhang was not a participant in Polymath8a, but Maynard was actively involved in the discussion in Polymath8b, which was enormously helpful given that of course he knew more than anyone else about his own work. A significant chunk of the discussion was between James Maynard and Terry Tao, trying to find ways to improve on the arguments that they had previously been working on independently and from slightly different perspectives.

Within days of Maynard posting his work on the arXiv, Polymath8b was obtaining useful finite bounds on H_2. Their work established that there are infinitely many triples of primes of width less than 500 000. (When I say 'width', I mean the distance between the smallest and largest, for example 11, 17, 19 is a triple of primes of width 8.) There was a sequence of bounds, each incrementally better than the last, all slightly less than 500 000.

As with Polymath8a, a range of people got involved in the discussion. James Maynard and Terry Tao were both very actively involved, drawing on their expertise about their new ideas and about the background sieve theory. They were joined by others, including Andrew Granville who is an expert on this sort of number theory (as well as being Maynard's mentor in Montreal). Along the way, there were various conversations about other related interesting problems: could the new work help with this question or with that? How does it connect with previous results about the primes? Making these links is a very fruitful avenue of investigation. Either the answer is known, and it's good to share that information, or it's a potential topic of research!

Polymathematicians were able to help with the computer calculations, thereby improving on Maynard's bounds, and so the numbers fell. In particular, Pace Nielsen, a mathematician at Brigham Young University in the US, worked on Maynard's Mathematica 'notebook', a computer file describing the program that he'd used. By developing these ideas further, and writing his own program, he was able to improve the bounds, although the time taken to do the calculations was something of a constraint: even on a computer it was taking days rather than hours! He shared his data via Tao's blog, and others joined in, poring over the data to see what they could learn. By the end of December 2013, Polymath8b had tentatively shown that there are infinitely many pairs of primes that differ by at most 272—and progress didn't stop there. (I say 'tentatively' because this was all part of an ongoing discussion, the work hadn't yet been carefully checked and written up for publication.)

The conversation continued through early 2014, with lots of detailed analysis of the arguments and the estimates and the computations, always looking for opportunities to improve on the bounds, and they had some success. On 9th February 2014, Terry Tao posted on his blog suggesting that it was time to start writing up the work for publication:

> Given the substantial progress made so far, it looks like we are close to the point where we should declare victory and write up the results (though we should take one last look to see if there is any room to improve the $H_1 \leq 270$ bounds).

By this stage Polymath8b was in a position to share a number of world records, and, even more importantly, the reasoning behind those bounds. But in fact the discussion on improving the argument was still very active, and continued to make progress.

By the middle of April 2014, around a year after Zhang's groundbreaking announcement that he could prove that there are infinitely many pairs of primes that differ by at most 70 000 000, Polymath8b had managed to improve that bound: they could show that there are infinitely many pairs of primes that differ by at most 246. Isn't that remarkable? I have no evidence that the Polymath collaborative approach led to mathematics that wouldn't have happened otherwise, but it is clear to me that it accelerated the process enormously, drawing together a group of interested people (the membership of the group varying over time), each with their own expertise to bring to bear on the problem.

As I write, this is the state of the art on this particular problem. I am not aware of any further progress, although it might be just round the corner—the announcement might come any time now, which makes writing this book an interesting challenge!

In parallel with the work on H_1, the bound described above, Polymath8b was making progress on bounding H_2 and other values, keeping track of larger clusters of primes that are very close together. The bound on H_2 was down below 400 000 by the end of 2013, and there were corresponding improvements to explicit bounds on H_3, H_4 and H_5.

There is another very successful aspect to the Polymath8b project. I mentioned in Chapter 7 that when you're stuck on a problem, a good way to make progress is to assume some other unproven result and see what you can deduce from that. Goldston, Pintz and Yıldırım did this back in 2005, when they assumed a suitably strong version of the Elliott–Halberstam Conjecture and deduced that there are infinitely many pairs of primes that differ by at most 16. Zhang's breakthrough was to give an *unconditional* proof of bounded gaps between primes—one that did not depend on any unproven conjectures—and Polymath, Maynard and Tao all continued to give unconditional arguments. But it's also possible to see what those arguments yield if we *do* assume the extra predictions about the distribution of the primes contained in the Elliott–Halberstam Conjecture. Maynard gave an unconditional proof that there are infinitely many pairs of primes that differ by at most 600, but in the same paper he also showed that if the Elliott–Halberstam Conjecture is true, then there are infinitely many pairs of primes that differ by at most 12, and infinitely many triples of primes of width at most 600 (that is, $H_1 \leq 12$ and $H_2 \leq 600$).

Dramatically, by the end of January 2014 Polymath8b was able to show that if a sufficiently strong conjecture about the distribution of the primes (the so-called *generalised Elliott–Halberstam Conjecture*) is true, then there are infinitely many pairs of primes that differ by at most 6. Just 6—which is particularly impressive because this is widely thought to be the best bound that it will be possible to prove using the current approach (thanks to the obstacle of the parity problem). This is extraordinarily close to the Twin Primes Conjecture, although of course it's not a complete proof for two reasons—it's not quite the desired bound of 2, and it relies on an unproven conjecture. But this is how progress in mathematics sometimes happens: inching forwards, getting a little closer, deepening our understanding all the time.

CHAPTER 16

Where next?

When a teacher gives a problem to a class of students, they usually know how to solve the problem. They usually know that the students have learned enough mathematical techniques to be able to solve the problem, and that the students have the mathematical sophistication and problem-solving skills to stand a good chance of solving the problem if they think hard about it for an appropriate amount of time.

Mathematical research is not like that. If you're tackling a problem that nobody in the history of humanity has solved before, then you have no idea what mathematical ideas and tools will go into a solution, whether you know them already, whether you need to develop them, whether you need to have a chance conversation with a colleague from another area of mathematics who can supply a tool from their branch of research, whether you need to wait a decade (or more) for someone else to discover a new approach. This is both exhilarating and terrifying! It also means that it is extremely hard, or perhaps impossible, to predict when mathematical breakthroughs will be made.

I believe that the Twin Primes Conjecture is true, and that one day we'll prove it. The preceding paragraph explains why I am not even going to guess when that might happen! It might be next year. It might be in ten years' time. It might be in thirty years' time, when a young person who's read this book has gone on to become a research mathematician tackling the problem .... To be completely honest, writing this book has been pretty nerve-wracking, because if someone proves the theorem before I get the manuscript to my publisher then I'm going to need to do some serious rewriting!

Maybe this is a good moment to talk briefly about some of the other recent progress on understanding the distribution of the primes. This is a hot topic at the moment, and I'm confidently expecting more exciting news over the coming months and years.

Large gaps between primes

We've thought lots in this book about small gaps between primes. What about large gaps between consecutive primes, do they occur? How large can they be?

In Chapter 8, I invited you to show that there are one hundred consecutive numbers that are *not* prime. I'm going to talk about a solution to this problem now, so this would be a good moment to think about it if you haven't already done so.

Thinking of one hundred consecutive composite numbers (*composite numbers* are those that are greater than 1 and not prime) might feel rather daunting. For me, this is one of those counterintuitive situations where making the problem harder can make it easier! (We saw this phenomenon with the Hardy-Littlewood circle method in Chapter 14.)

What if instead I invite you to think of one hundred consecutive numbers bigger than 100 with the property that the first is a multiple of 2, the second a multiple of 3, the third a multiple of 4, the fourth a multiple of 5, and so on? That would answer the question, because a multiple of 2 that's bigger than 100 cannot be prime, similarly a multiple of 3 that's bigger than 100, and so on. So that would yield one hundred consecutive composite numbers. Instinctively it feels like a harder problem, because we're imposing extra constraints (we want one hundred consecutive composite numbers of a particular form, not just any old hundred consecutive composite numbers), but these extra constraints are actually useful for helping us to think of a suitable example.

We can use an idea inspired by the one Euclid used in his proof that there are infinitely many primes.

Let's multiply together 2, 3, 4, 5, 6, and so on, all the way up to 100. We often write this as 100!—we read it as '100 factorial'. It simply means $100 \times 99 \times 98 \times \cdots \times 4 \times 3 \times 2 \times 1$.

Now $100! + 2$ is a multiple of 2 (because certainly 100! is—we included 2 in the product), and $100! + 3$ is a multiple of 3, and $100! + 4$ is a multiple of 4, and so on, all the way up to $100! + 100$, which must be a multiple of 100. So we have a whole bunch of consecutive composite numbers!

Actually, we have only ninety-nine consecutive composite numbers. Knowing that $100! + 1$ is a multiple of 1 doesn't help a lot. But we can easily fix this, by instead considering 101! + 2, 101! + 3, 101! + 4, 101! + 5, ..., 101! + 100, 101! + 101—and these really are one hundred consecutive composite numbers.

Better still, we can extend the same idea to find as many consecutive composite numbers as we like. You'd like one million consecutive composite numbers? No problem, 1 000 001! + 2, 1 000 001! + 3, 1 000 001! + 4, 1 000 001! + 5, ..., 1 000 001! + 1 000 000, 1 000 001! + 1 000 001 will do the job.

What was the point of all this? Well, we wanted to know whether we can have large gaps between consecutive primes, and we have shown that these gaps can be *arbitrarily large*, because we can find as many consecutive composite numbers as we like.

You won't by now be surprised to learn that this is not the end of the story. The numbers 101! + 2, 101! + 3 and so on are *huge*. Properly enormous. Do we really need to go so far to find a block of one hundred consecutive composite numbers? It was very convenient to go so far, because it was easy to see that 101! + 2 and so on really are composite, and this helped us to answer the question 'Are there one hundred consecutive composite numbers?' in a concise way. But it would be interesting to have a clearer sense of when these blocks of non-primes appear.

We saw in Chapter 8 that the Prime Number Theorem, which estimates the number of primes up to any given point, tells us that on average the gap between consecutive primes of size around x is about $\log x$. When would we expect the *average* gap between primes to be 100? Well, between primes somewhere around e^{100}, which is substantially smaller than 101! (because to find e^{100} we multiply $e \approx 2.71$ by itself 100 times, so have a product of 100 numbers all less than 3, whereas in $101! = 101 \times 100 \times 99 \times \cdots \times 3 \times 2$ we have a product of 100 numbers almost all of which are bigger, much bigger, than 3). So actually we expect the first block of one hundred consecutive composite numbers to occur a lot sooner than starting at 101! + 2—but proving this is much more difficult than observing that 101! + 2 up to 101! + 101 are all composite!

It turns out that it is possible to say much more about large gaps between primes. In August 2014, on successive dates two new preprints appeared on the arXiv website on this subject. On 20th August 2014, Kevin Ford, Ben Green, Sergei Konyagin and Terence Tao posted an article on 'Large gaps between consecutive prime numbers'. (We have already met Ben Green and Terence Tao in earlier chapters. Kevin Ford is a mathematician at the University of Illinois at Urbana-Champaign, while Sergei Konyagin is a mathematician based at the Steklov Mathematical Institute of the Russian Academy of Sciences in Moscow.) The following day, James Maynard uploaded his work on 'Large gaps between primes'. The authors had, in different ways, managed to show that the gap between consecutive primes up to x can be rather larger than the average (giving a lower bound that is more complicated than I want to write here—like many such bounds in analytic number theory it involves many logarithms!). Intriguingly, Maynard had managed to take his work on *small* gaps between primes, and find a way to adapt it to reveal information about *large* gaps between primes! Ford, Green, Konyagin and Tao used a different approach, drawing on the work of Green and Tao on arithmetic progressions in the primes (which we met in Chapter 6).

On Google+, Tim Gowers highlighted the coincidence of these two independent pieces of work appearing at the same time. Ben Green, one of the co-authors of the first paper, noted

> 'by convention this absolutely counts as independent discovery with equal credit, even if the precise dates are slightly different. It would have counted as such even if Maynard had only had rough notes* on his approach at the time we arxived our paper, though as it happens his paper was already complete. I actually consider it a little unfortunate that FGKT put our paper on the arxiv before Maynard, but that resulted from us not having any idea about his work. Pure mathematics is a bit different from some other branches of science in this regard!
>
> *such that the veracity of the argument could be quickly verified by experts.'

Mathematicians tend to be civilised about such matters: as we've already seen, Maynard and Tao came to an amicable agreement about how to handle their having proved similar results at a similar time. In that case, they had used similar arguments too, whereas the approach of Ford, Green, Konyagin and Tao on long gaps between primes was different from that used by Maynard. In both examples, the facts are in the open. It is very clear to anyone who chooses to look that there were two people/groups who independently proved the result at a similar time, and that is surely more helpful than worrying about whether one person/group posted their work online a few hours earlier than another. As Gowers observed on the Google+ discussion,

> One benefit . . . is that it is a genuinely useful piece of information to know that a result has been proved independently by more than one person. It indicates (though does not prove beyond all doubt) that certain ideas were in the air rather than the result being a complete bolt from the blue.

When the two groups of authors became aware of the other piece of work, they realised that by pooling their ideas they could go even further, obtaining an improved lower bound. In December 2014, Ford, Green, Konyagin, Maynard and Tao posted a first draft of their preprint, explaining their new argument. This is not the end of the story, because, as with the Twin Primes Conjecture, there are reasons to believe that more is true than mathematicians can currently prove, but it's an exciting step in the right direction. And while this collaboration is not so large as a Polymath project, it does nicely highlight the value of mathematicians sharing ideas, to make more progress than any of them would have managed individually (at least in the same timescale).

One interesting aspect of the initial two papers of Ford, Green, Konyagin, Tao and of Maynard was that they answered a longstanding question of Paul Erdős. Erdős was truly a prolific mathematician: a prolific answerer of questions, but also a prolific asker of questions. He often offered monetary rewards for answers to questions, with the amount of money promised indicating how hard he thought the question was. Back in 1934, when he was in his early twenties, Erdős himself had proved a certain lower bound on the gap between consecutive primes up to a given point, and this was improved a little in 1938 by the British mathematician Robert Rankin (1915–2001). Erdős offered, 'perhaps somewhat rashly', \$10 000 for a proof of the bound that was finally obtained in 2014 by Ford, Green, Konyagin, Tao and, independently, by Maynard—an unusually large amount for him to offer, presumably reflecting how hard he perceived the problem to be. Since Erdős died in 1996, his friend and collaborator Ron Graham (an American mathematician) has been honouring his promises of financial prizes, and so Ford, Green, Konyagin and Tao between them received half the money, and Maynard the other half (corresponding to the two papers that independently solved the problem).

As so often happens in mathematics, the significance of a proof of a result is not only the theorem itself, but also the creation of new strategies for tackling other problems. Maynard illustrated that beautifully when he took his ideas for proving results about small gaps between primes and developed them to prove theorems about *large* gaps between primes! Others have been quick to scrutinise the work of Zhang, Polymath, Maynard, Tao and others, to try to understand which problems might be worth tackling using their new ideas. Some of these are other problems about the distribution of the primes, some are analogous problems in settings that are generalisations of the integers (such as the Gaussian integers that we met in Chapter 12), some are problems where it is a surprise that these ideas are relevant. The recent breakthroughs have given a major boost to this branch of analytic number theory, and this is surely a sign that there are more breakthroughs to come.

Is Polymath the future?

In a hundred years' time, will all mathematics be done via large, public collaborations such as Polymath? I don't think so—but I do think that such collaborations are going to become more common. It seems clear that some mathematical projects lend themselves to a Polymath-style approach, but not all. Why was Polymath so spectacularly successful on bounded gaps between primes? The notoriety of the problem, helped by the fact that it's relatively easy to describe the question, meant that many people wanted to be part of the project—they were excited about the opportunity to be part of making history. However, this was not the only feature that meant it worked well for Polymath. After all, although the problem is not hard to explain, the technicalities of the recent papers on the subject placed a hurdle to be overcome in order to participate. However, some people were so motivated by the problem that they were willing to put in the homework in order to get over that hurdle, and, crucially, others were able to contribute in meaningful ways even without following the technical details, perhaps most notably by providing computing expertise. The people who wanted to think about the theoretical bounds were not necessarily the same as the people who were good at getting computers to run lengthy calculations accurately and efficiently, but there was enough common language between those groups that the collaboration worked very effectively.

Sometimes, there is nothing to beat an individual, or a small group of colleagues who know each other, sitting and thinking very hard about a problem for a prolonged period of time, occasionally taking time away to work on other questions and allowing their subconscious to take over. Polymath will not work for every question—or indeed for every mathematician, some of whom prefer to work quietly in private. But the arrival of the internet, and with it Polymath, means that there are exciting opportunities for continuing to develop these new ways of collaborating, and for seeking to understand further which kinds of projects are well suited to such collaborations.

Understanding the relative strengths and weaknesses of the Polymath approach compared to a more conventional collaboration is itself a subject of research. Ursula Martin and Alison Pease, two British academics who work in mathematics and computer science, are investigating this area, seeking to understand the role of collaboration in mathematical research. They have studied the famous collaboration of Hardy and Littlewood in the first half of the twentieth century, contrasting it with the Polymath collaborations of the twenty-first century. They write

> *polymath* succeeded in opening up the process of mathematical discovery to a wide audience from a variety of backgrounds, and the ground rules, and the involvement of senior figures on the discussions, reinforced in a very open and public way expectations of how mathematical collaboration should be conducted.

but go on to conclude

> A number of participants and possible participants remarked on their distaste for the fast pace of *polymath* discussions, which requires people to work quickly and not mind making mistakes in public, and wondered whether this is really a necessary condition for collaboration, or merely a way for people who are comfortable with this to gain status in the community. One feels that G H Hardy, who hated the highly competitive nineteenth century Cambridge mathematical tripos, and was well known for his reticence with strangers, would have agreed.

There is still much to be learned about carrying out large-scale mathematical collaborations such as those envisaged by Tim Gowers when he proposed Polymath, but the experiences of those who participated in the first few Polymath projects (and of those who watched from the sidelines but chose not to participate) will no doubt help to inform the collaborations of the future.

Some of the participants in Polymath8 contributed to a retrospective article published in the Newsletter of the European Mathematical Society, in which they reflected on their experiences. Terry Tao describes in some detail how the project began and unfolded, and how he found the experience of being involved. He concludes his account by saying

> All in all, it was an exhausting and unpredictable experience but also a highly thrilling and rewarding one.

In total, ten individuals contributed their personal perspectives to this article, between them covering a range of career stages and a range of levels of involvement with Polymath8. Two recurring themes quickly arise (as Ursula Martin and Alison Pease noted in their work): the pace of the project and level of commitment required to keep up, and the experience of posting in public ideas that subsequently turned out to be incorrect. For example, James Maynard wrote

> I was surprised at how much time I ended up devoting to the Polymath project. This was partly because the nature of the project was so compelling—there were clear numerical metrics of 'progress' and always several possible ways of obtaining small improvements, which was continually encouraging. The general enthusiasm amongst the participants (and others outside of the project) also encouraged me to get more and more involved in the project.

Andrew Sutherland agreed:

> Like others, I was surprised by how much time I ended up devoting to the project. The initially furious pace of improvements and the public nature of the project made a very addictive combination and I wound up spending most of that summer working on it. This meant delaying other work but my collaborators on other projects were very supportive. I certainly do not begrudge the time I devoted to the Polymath8 effort; it was a unique opportunity and I'm glad I participated.

The time commitment was a source of concern for some participants, partly early career academics for whom future job applications have to be a consideration. Will hiring committees take into account contributions to a collaboration such as Polymath?

Pace Nielsen wrote

> ...a large number of mathematicians I know commented (in personal communications to me) on the fact that they were 'impressed with my bravery' in participating. It caught me off guard that so many people had been following the project online and that all of my comments (including my mistakes) were open to such a wide readership. I believe it is important to consider this issue before deciding to participate in a public project. Some of the mistakes I made would never have seen the light of day in a standard mathematical partnership. However, any collaboration relies on the ability for the participants to share ideas freely, even the 'dumb' ones.

Making public and preserving *all* the ideas, not just the ones that contributed to the final paper, and certainly not just the correct ones, is one of the distinctive aspects of Polymath. It is a disconcerting experience for many of the participants: mathematicians are often by nature inclined to work very hard to get things right, to be precise, not to make sloppy statements that are incorrect—but it is by bouncing around ideas and refining sloppy statements that progress is made, and it is misleading to students if they deduce from their textbooks that mathematics always arrives in a refined form that is precisely correct! Gergely Harcos remarked

> At one point I embarrassed myself by posting several different 'proofs' to an improved inequality that I conjectured, only to find out later that the claim was false. All this is recorded and preserved in the blog but I do not regret it as it was honest and reflects the way mathematics is done. We try and we often fail.

There were many people whose mathematical careers were affected by Polymath8, not only the active participants. Andrew Gibson, an undergraduate at the University of Memphis, wrote

> ... reading the posts and following the 'leader-board' felt a lot like an academic spectator sport. It was surreal, a bit like watching a piece of history as it occurred. It made the mathematics feel much more alive and social, rather than just coming from a textbook. I don't think us undergrads often get the chance to peak [sic] behind closed doors and watch professional mathematicians 'in the wild' like this so, from a career standpoint, it was illuminating.

I am sure that there will be future Polymath projects, and I expect that at least some of them will turn out to be mathematically successful, although perhaps few will reach the levels of public prominence that Polymath8 managed. However, Polymath8 itself is not exhausted, and there is still more to be learned by looking at the records of the project as it progressed.

As I mentioned earlier in this book, some of the key protagonists in this story have been acknowledged by the mathematical community for their work on bounded gaps between primes: Zhang and Maynard, in particular, have rightly been honoured with prizes. It is less clear whether the mathematical community will feel comfortable awarding a prize to Polymath, to a massive and (currently) unorthodox collaboration of a kind previously unknown in mathematics. But the role of the Polymathematicians in understanding bounded gaps between primes seems to me to be very significant, and even if the judging panels for mathematical prizes do not feel that this particular aspect of Polymath's work is worthy of a prize, surely at some point in the future they are going to have to award a prize to a Polymath project. That moment will be a significant landmark in the Polymath journey.

The future

I cannot predict what will come next, what the next breakthroughs will be. But I am certain that there will be breakthroughs in our understanding of the primes, sometimes small results that chip away, sometimes big hammer blows that break the whole rock open and leave a whole pile of shards to study. There has never been a more exciting time to study prime numbers! And, for now, the quest to close the gap continues . . .

FURTHER READING

There are many excellent books and websites on the topics discussed in this book, and what follows is not supposed to be a complete list. Rather, I have selected a few personal favourites together with some of the source material for the recent work on prime numbers (but what follows is not a complete list of references). They are organised into sections by topic, and within sections are roughly arranged in the order in which they are relevant for this book.

All URLs are correct as of January 2017.

General mathematics books and websites

- The MacTutor History of Mathematics archive http://www-history.mcs.st-and.ac.uk/

  This website is run by John O'Connor and Edmund Robertson at the University of St Andrews. It is an absolutely invaluable collection of biographies of mathematicians, as well as further information on the history of mathematics. It was very helpful to me in writing this book, and you will find that it contains much more biographical information on most of the mathematicians that I have mentioned, together with many pictures.

- G.H. Hardy, *A Mathematician's Apology*, Cambridge University Press, 2012

  This book, written in 1940, is an extraordinary insight into what it is to do mathematics. Written by Hardy late in his career, it has an elegiac quality. It is full of quotes that other mathematicians have subsequently used to describe what they do. A must read.

- Plus Magazine https://plus.maths.org/

  Plus Magazine is part of the Millennium Mathematics Project at the University of Cambridge. They have lots of very accessible and engaging articles, features and interviews relating to mathematical research of the past and present.

Number theory books and websites

- Animated factorisation diagrams http://www.datapointed.net/visualizations/math/factorization/animated-diagrams/
  This website was the inspiration for the dotty pictures at the start of Chapter 2. It shows them in an animated way, and is beautiful to watch. Check your understanding by predicting the diagram for a particular number and then watch through till the film reaches that number!

- Simon Singh, *Fermat's Last Theorem*, Fourth Estate, 1997 and 2002 new edition, also BBC Horizon television documentary
  Shortly after Fermat's Last Theorem was proved, Simon Singh and John Lynch made a television documentary about it, for the BBC Horizon series. It is still available (at least in the UK) via the BBC iPlayer website. If you can, you should watch it: it is outstandingly good. Singh went on to write a book, which is similarly excellent.

- H. Davenport, *The Higher Arithmetic*, Eighth edition, Cambridge University Press, 2008
  The 'higher arithmetic' is an old-fashioned name for number theory, and the book is an introduction to the subject. This is in some sense a mathematics textbook: it goes into the details, and includes exercises. However, it does not read like a textbook—it is beautifully written, and a real favourite of mine. Davenport also wrote the less-than-snappily titled *Analytic methods for Diophantine Equations and Diophantine Inequalities* (Cambridge University Press, 2005), which is one of my favourite introductions to the Hardy–Littlewood circle method, but this is significantly more technical than *The Higher Arithmetic*.

- Marcus du Sautoy, *The Music of the Primes: Why an Unsolved Problem in Mathematics Matters*, HarperPerennial, 2004
  A gentle introduction to the Riemann Hypothesis and related aspects of number theory, with plenty about the history and the mathematicians involved in the story.

Polymath

- Michael Nielsen *Doing science online* http://michaelnielsen.org/blog/doing-science-online/, 26 January 2009
  A blog post in which Michael Nielsen describes his thoughts on doing science online, for example, highlighting blogging amongst mathematicians. Links within this blog post prompted Tim Gowers to go public with his ideas about Polymath.

- Tim Gowers *Is massively collaborative mathematics possible?* https://gowers.wordpress.com/2009/01/27/is-massively-collaborative-mathematics-possible/, 27 January 2009

The blog post in which Tim Gowers first proposes the idea of Polymath, and set out the rules for the project.

- Tim Gowers *A combinatorial approach to density Hales–Jewett* https://gowers.wordpress.com/2009/02/01/a-combinatorial-approach-to-density-hales-jewett/, 1 February 2009

  The blog post in which Tim Gowers starts the first Polymath project. I am not going to list all the subsequent blog posts relating to this Polymath project or the subsequent projects! But this is a good place to start if you want to know more about what the first Polymath project was about. More generally, you will find many Polymath-related posts on Gowers's blog https://gowers.wordpress.com/.

- Terry Tao *IMO 2009 Q6 mini-polymath project: impressions, reflections, analysis* https://terrytao.wordpress.com/2009/07/22/imo-2009-q6-mini-polymath-project-impressions-reflections-analysis/, 22 July 2009

  A blog post in which Terry Tao reflects on the mini-Polymath project that he created to tackle an International Mathematics Olympiad question, and invites others to comment on the project.

- *The polymath blog* http://polymathprojects.org

  A blog run by Tim Gowers, Gil Kalai, Michael Nielsen and Terry Tao, to host Polymath projects.

- *The Erdős discrepancy problem* http://michaelnielsen.org/polymath1/index.php?title=The_Erd%C5%91s_discrepancy_problem

  The project wiki for the Polymath5 project on the Erdős discrepancy problem.

- Terry Tao *Sign patterns of the Mobius and Liouville functions* https://terrytao.wordpress.com/2015/09/06/sign-patterns-of-the-mobius-and-liouville-functions/, 6 September 2015

  The blog post in which Tao described his work with Kaisa Matomäki and Maksym Radziwiłł. This prompted a comment from Uwe Stroinski (https://terrytao.wordpress.com/2015/09/06/sign-patterns-of-the-mobius-and-liouville-functions/#comment-459021) that helped Tao to link this work to the Erdős discrepancy problem.

- Terry Tao *The Erdos discrepancy problem via the Elliott conjecture* https://terrytao.wordpress.com/2015/09/11/the-erdos-discrepancy-problem-via-the-elliott-conjecture/, 11 September 2015

  The blog post in which Tao drew on work of Polymath5 to show that a proving a certain conjecture, the non-asymptotic Elliott conjecture, would give a solution to the Erdős discrepancy problem.

- Terry Tao *The logarithmically averaged Chowla and Elliott conjectures for two-point correlations; the Erdos discrepancy problem* https://terrytao.

wordpress.com/2015/09/18/the-logarithmically-averaged-chowla-and-elliott-conjectures-for-two-point-correlations-the-erdos-discrepancy-problem/, 18 September 2015

The blog post in which Tao announces two new papers, including his solution of the Erdős discrepancy problem, and outlines his work.

- Terence Tao *The Erdős discrepancy problem* in *Discrete Analysis*, 2016:1, 27pp, available at http://discreteanalysisjournal.com/article/609-the-erdos-discrepancy-problem

  The paper in which Tao answers the Erdős discrepancy problem. Note that the journal website includes an editorial description of the paper, and links to the full paper on the arXiv (this is the format of this journal).

- Tim Gowers *EDP28 — problem solved by Terence Tao!* https://gowers.wordpress.com/2015/09/20/edp28-problem-solved-by-terence-tao/, 20 September 2015

  In this blog post Tim Gowers reflects on the first few Polymath projects, and describes why he regards Polymath5 as a success despite the project having not solved the problem it ostensibly set out to solve.

- Erica Klarreich *A Magical Answer to an 80-Year-Old Puzzle* in *Quanta Magazine*, October 2015, https://www.quantamagazine.org/20151001-tao-erdos-discrepancy-problem/

  In this article, aimed at a non-expert audience, Erica Klarreich describes the Erdős discrepancy problem and its solution by Tao. This would be a great place to start reading about the problem if you don't already know about it.

- Ursula Martin and Alison Pease *Hardy, Littlewood and polymath* in E. Davis, P.J. Davis (eds.), *Mathematics, Substance and Surmise*, Springer 2015, pp. 9–23, DOI 10.1007/978-3-319-21473-3_2, http://www.springer.com/us/book/9783319214726

  In this book chapter, Ursula Martin and Alison Pease compare the famous collaboration of Hardy and Littlewood in the first part of the twentieth century with the Polymath collaboration in the twenty-first century.

Bounded gaps between primes—the start and Yitang Zhang

- D.A. Goldston, J. Pintz, C.Y. Yıldırım *Primes and tuples I* in *Annals of Mathematics*, vol. 170 (2009), pp. 819–862, abstract (and full paper for subscribers) available at http://annals.math.princeton.edu/2009/170-2/p10 and preprint available at https://arxiv.org/abs/math/0508185

  The paper in which Daniel Goldston, János Pintz and Cem Yıldırım show that if the Elliott–Halberstam Conjecture is true, then there are infinitely many pairs of primes that differ by at most 16.

- Yitang Zhang *Bounded gaps between primes* in *Annals of Mathematics*, vol. 179 (2014), pp. 1121–1174, abstract (and full paper for subscribers) available at http://annals.math.princeton.edu/2014/179-3/p07

In this paper Yitang Zhang shows that there are infinitely many pairs of primes that differ by at most 70 000 000.

- Erica Klarreich *Unheralded Mathematician Bridges the Prime Gap* in *Quanta Magazine*, May 2013, https://www.quantamagazine.org/20130519-unheralded-mathematician-bridges-the-prime-gap/
  In this article, aimed at a non-expert audience, Erica Klarreich describes the story of Zhang's momentous breakthrough.

- *Philosophy behind Yitang Zhang's work on the Twin Primes Conjecture* http://mathoverflow.net/questions/131185/philosophy-behind-yitang-zhangs-work-on-the-twin-primes-conjecture/, started in May 2013
  A question-and-answer format in which mathematicians start to discuss what was new about Zhang's work and how far it would be possible to improve the bounds.

- Scott Morrison *I just can't resist: there are infinitely many pairs of primes at most 59470640 apart* https://sbseminar.wordpress.com/2013/05/30/i-just-cant-resist-there-are-infinitely-many-pairs-of-primes-at-most-59470640-apart/, 30 May 2013
  The blog post in which Scott Morrison starts the sport of briefly holding the world record for the best known bound by improving on Zhang's argument.

Bounded gaps between primes—the Polymath project and James Maynard

- *Bounded gaps between primes* http://michaelnielsen.org/polymath1/index.php?title=Bounded_gaps_between_primes
  The homepage for the Polymath8 project.

- *Timeline of prime gap bounds* http://michaelnielsen.org/polymath1/index.php?title=Timeline_of_prime_gap_bounds
  The wiki 'league table' recording the best known results on bounded gaps between primes.

- *Polymath8 grant acknowledgements* http://michaelnielsen.org/polymath1/index.php?title=Polymath8_grant_acknowledgments
  A page on which Polymath8 participants record their contributions and acknowledge their receipt of grant funding as appropriate.

- Terry Tao's blog, the Polymath section https://terrytao.wordpress.com/category/question/polymath/
  Much of the discussion for Polymath8 happened on Tao's blog, but it would be impractical to link to every relevant post here, so I shall just encourage you to look through this section of his blog

- D.H.J. Polymath *New equidistribution estimates of Zhang type* preprint available at https://arxiv.org/abs/1402.0811

One of the papers recording the progress made by Polymath8, first submitted to the arXiv on 4th February 2014.

- D.H.J. Polymath *Variants of the Selberg sieve, and bounded intervals containing many primes* preprint available at https://arxiv.org/abs/1407.4897.

  The paper in which, amongst other things, Polymath shows that there are infinitely many pairs of primes that differ by at most 246. The paper was first submitted to the arXiv on 18th July 2014.

- *Narrow admissible tuples* http://math.mit.edu/~primegaps/

  This library of admissible sets is maintained by Andrew Sutherland, with contributions from others too, and was and continues to be a useful resource for Polymathematicians and others working on bounded gaps between primes. It is the source of the examples of admissible sets that I gave in Chapters 7 and 13 (the latter was used by Maynard, and is available at http://math.mit.edu/~primegaps/tuples/admissible_105_600.txt).

- James Maynard, *Small gaps between primes* in *Annals of Mathematics*, vol. 181 (2015), pp. 383–413, abstract (and full paper for subscribers) available at http://annals.math.princeton.edu/2015/181-1/p07 and preprint available at https://arxiv.org/abs/1311.4600

  In this paper, Maynard introduces new ideas and proves that there are infinitely many pairs of primes that differ by at most 600 (amongst other results).

- D.H.J. Polymath, *The "Bounded Gaps between Primes" Polymath Project: A Retrospective Analysis*, Newsletter of the European Mathematical Society, December 2014, Issue 94, pp. 13–23, available at http://www.ems-ph.org/journals/newsletter/pdf/2014-12-94.pdf

  In this article, ten Polymath8 participants give their own personal reflections on the experience of being involved with, or of observing, the project. There is also a helpful bibliography with many references to related work on bounded gaps between primes.

- Terry Tao *Open question: The parity problem in sieve theory* https://terrytao.wordpress.com/2007/06/05/open-question-the-parity-problem-in-sieve-theory/, 5 June 2007

  A blog post in which Tao introduces some ideas from sieve theory, the obstacles presented by the parity problem. If you are interested in the parity problem then you might also like another blog post by Tao, *The parity problem obstruction for the binary Goldbach problem with bounded error* https://terrytao.wordpress.com/2014/07/09/the-parity-problem-obstruction-for-the-binary-goldbach-problem-with-bounded-error/, 9 July 2014.

- *The Pursuit of Beauty* in *The New Yorker*, 2 February 2015, available at http://www.newyorker.com/magazine/2015/02/02/pursuit-beauty

  A profile of Yitang Zhang and a description of the recent developments on understanding bounded gaps between primes.

- *Together and Alone, Closing the Prime Gap* in *Quanta Magazine*, November 2013, https://www.quantamagazine.org/20131119-together-and-alone-closing-the-prime-gap/

  Erica Klarreich describes the 2013 breakthroughs on bounded gaps between primes, including the work by James Maynard.

- Andrew Granville *Primes in intervals of bounded length* Bulletin of the American Mathematical Society, vol 52, no 2, April 2015, pp. 171–222, and available at http://www.dms.umontreal.ca/~andrew/PDF/Bulletin14.pdf

  An article in which Andrew Granville describes the work of Zhang, Maynard, Tao and Polymath in some detail and with great clarity. Granville dedicated the article 'To Yitang Zhang, for showing that one can, no matter what'.

- Ben Green *Bounded gaps between primes* http://arxiv.org/abs/1402.4849

  These are technical notes that Ben Green prepared to explain the work of Zhang and others to a mathematical audience.

More on prime numbers

- *Goldbach Conjectures* https://xkcd.com/1310/

  A not-entirely-serious take on Goldbach's Conjecture and the weak Goldbach Conjecture.

- H.A. Helfgott *The ternary Goldbach conjecture is true*, preprint available at https://arxiv.org/abs/1312.7748

  The paper in which Helfgott closes the gap to prove the weak Goldbach conjecture, drawing on the computer-generated data from

- H.A. Helfgott and David J. Platt *Numerical Verification of the Ternary Goldbach Conjecture up to* $8.875 \cdot 10^{30}$ in *Experimental Mathematics* vol. 22 (2013), pp. 406–409 at http://www.tandfonline.com/doi/full/10.1080/10586458.2013.831742, preprint available at http://arxiv.org/abs/1305.3062

- Ben Green and Terence Tao *The primes contain arbitrarily long arithmetic progressions* in *Annals of Mathematics*, vol. 167 (2008), pp. 481–547, available at http://annals.math.princeton.edu/2008/167-2/p03 and preprint available at https://arxiv.org/abs/math/0404188

  The paper in which Green and Tao prove their famous theorem that there are 50 evenly spaced prime numbers, and 500, and in fact arbitrarily many.

- Andrew Granville and Greg Martin *Prime Number Races* in *The Americal Mathematical Monthly*, vol. 113, 2006, pp. 1–33, and available at http://www.dms.umontreal.ca/~andrew/PDF/PrimeRace.pdf

  In this article, Andrew Granville and Greg Martin describe, admirably clearly and with lots of detail, the recent work on the so-called prime number races. Pick your favourite number, and count the number of primes up to that point that are one more than a multiple of 4 and the number that are three more than a multiple of 4. Much of the time, it seems that the latter are in the lead, that there are more primes that are three more than a multiple of 4. But not always. So what's going on, and can this be generalised?

- Kevin Ford, Ben Green, Sergei Konyagin, Terence Tao *Large gaps between consecutive prime numbers* in *Annals of Mathematics*, vol. 183 (2016), pp. 935–974, abstract (and full paper for subscribers) available at http://annals.math.princeton.edu/2016/183-3/p04 and preprint available at http://arxiv.org/abs/1408.4505

  The paper in which Ford, Green, Konyagin and Tao answer Erdős's famous question about large gaps between prime numbers.

- James Maynard *Large gaps between primes* in Annals of Mathematics, vol. 183 (2016), pp915–933, abstract (and full paper for subscribers) available at http://annals.math.princeton.edu/2016/183-3/p03 and preprint available at https://arxiv.org/abs/1408.5110

  The paper in which Maynard gives his answer to Erdős's question about large gaps between prime numbers.

- Kevin Ford, Ben Green, Sergei Konyagin, James Maynard, Terence Tao *Long gaps between primes*, preprint available at http://arxiv.org/abs/1412.5029

  The paper in which Ford, Green, Konyagin and Tao come together with Maynard to improve on their previous lower bounds for large gaps between consecutive primes.

INDEX